CISM Courses and Lectures

The series presents lecture notes, monographs, edited works and proceedings in the field of Mechanics, Engineering, Computer Science and Applied Mathematics.
Purpose of the series is to make known in the international scientific and technical community results obtained in some of the activities organized by CISM, the International Centre for Mechanical Sciences.

International Centre for Mechanical Sciences

Courses and Lectures Vol. 564

For further volumes:
www.springer.com/series/76

Antonello Provenzale · Elisa Palazzi · Klaus Fraedrich
Editors

The Fluid Dynamics of Climate

Springer

Editors

Antonello Provenzale
Institute of Geosciences and Earth Resources, CNR
Pisa, Italy

Elisa Palazzi
Institute of Atmospheric Sciences and Climate, CNR
Torino, Italy

Klaus Fraedrich
Max Planck Institute for Metereology
Hamburg, Germany

ISSN 0254-1971
ISBN 978-3-7091-1891-7 ISBN 978-3-7091-1893-1 (eBook)
DOI 10.1007/ 978-3-7091-1893-1
Springer Wien Heidelberg New York Dordrecht London

All contributions have been typeset by the authors

Printed in Italy

Printed on acid-free paper

Springer is part of Springer Science+Business Media (www.springer.com)

PREFACE

Climate dynamics offers some of the most intriguing scientific problems in science, as well as a set of applied issues of central importance, such as the definition of mitigation and adaptation strategies, the assessment of the potential risks associated with climate change (droughts, floods, extreme events, sea level rise) and the social, economic and geopolitical implications of global warming.

Many of the components of the climate system are in fluid state, such as the atmosphere, the hydrosphere and the cryosphere. As an evolution of the well-established discipline of geophysical fluid dynamics, founded more than fifty years ago, the emerging theme of "climatic fluid dynamics" is at the heart of the efforts devoted to understanding and modeling the climate system.

On August 26-30, 2013, the International Center for Mechanical Sciences (CISM) organized in Udine the course "The Fluid Dynamics of Climate", directed by Klaus Fraedrich from the University of Hamburg and Antonello Provenzale from the National Research Council of Italy. The objective of the course was to make students and researchers with a general background in fluid dynamics familiar with the fluid aspects of the climate system. The course brought together contributions from diverse fields of the physical, mathematical and engineering sciences. The addressed audience was composed of doctorate students, postdocs and researchers working on different aspects of atmospheric, oceanic and environmental fluid dynamics, as well as researchers interested in quantitatively understanding how fluid dynamics can be applied to the climate system, and climate scientists willing to gain a deeper insight into the fluid mechanics underlying climate processes.

This volume collects the lecture notes of the course. First, the contribution by Henk Dijkstra addresses the dynamical systems approach to ocean and climate dynamics. Specific topics include ENSO and the Atlantic Multidecadal Oscillation. The contribution by Claudia Pasquero then illustrates some of the convective processes taking place in the ocean and the atmosphere, crucial to the workings of the climate system. The third contribution, by Miklós Vincze and Imre Jánosi, discusses laboratory experiments on large-scale geophysical flows, that provide direct insight into climatic processes.

The volume continues with three more specific contributions: Tímea Haszpra and Támas Tél discuss the Lagrangian, particle-based description of large-scale atmospheric motions, adopting a dynamical systems approach. The contribution by Annalisa Bracco and co-authors addresses the crucial problem of parameter optimization in climate models and some of the open issues in model intercomparison projects. The contribution of Klaus Fraedrich, Isabella Bordi and Xiuhua Zhu considers climate dynamics at global scale, specifically looking at resilience, hysteresis and the attribution of change.

The final part of the volume contains two contributions on different aspects of the water cycle: Elisa Palazzi and Antonello Provenzale discuss some general aspects of the water cycle in the climate system, considering large-scale moisture transport, soil-atmosphere interactions and climate downscaling. The contribution by Klaus Fraedrich and coworkers focuses on climate dynamics at the watershed scale, addressing the rainfall-runoff chain.

All together, these notes provide a glimpse of how deeply fluid dynamics is rooted in the study of climate, and how fertile the approach based on using fluid dynamical concepts in climate science can be. We hope the readers will enjoy these notes as much as we enjoyed the course and the lectures.

Clearly, neither the course nor the volume could have been realized without the generous support of the International Center for Mechanical Sciences, which we warmly thank. We are also grateful to the Editors of the CISM book series for their patience in waiting for the notes to be completed.

Antonello Provenzale, Elisa Palazzi and Klaus Fraedrich

CONTENTS

Understanding climate variability using dynamical systems theory

Henk A. Dijkstra*

* Institute for Marine and Atmospheric research Utrecht, Department of Physics and Astronomy, Utrecht University, Utrecht, The Netherlands

Abstract In this chapter, an introduction will be given on how to use the theory and methods of stochastic dynamical systems theory to understand phenomena of climate variability. We will restrict the application of this theory to very basic aspects of the El Niño/Southern Oscillation and the Atlantic Multidecadal Oscillation.

1 Introduction

Although climate scientist's views on the climate system probably greatly differ, most would admit that it is a system displaying very complex spatio-temporal variability in many of its components such as the atmosphere, the hydrosphere (including the oceans), the cryosphere, the biosphere and the lithosphere.

In a report to the NASA Advisory Council, Bretherton (1988) presented a sketch of the Earth System components and their interactions. The original figure, sometimes referred to as the 'horrendogram' of the climate system, and its simplification shown in Fig. 1 are certainly useful in recognizing many of the subcomponents of the climate system and identifying important processes. The figure also provides a basis to understand the transfer of properties (such as energy and mass) that are exchanged between these different subsystems. Examples of such interactions and associated fluxes are usually referred to as the energy cycle, the hydrological cycle and several biogeochemical cycles (for example, the carbon cycle).

To understand climate variability, it is important to realize that different characteristic time scales are introduced into the climate system by the different processes in the subsystems. One way of looking at these time scales is to perturb the specific subsystem out of an equilibrium and then monitor how long it takes for it to reach equilibrium again. Such characteristic (or response) time scales of atmospheric processes range from a few

A. Provenzale et al. (Eds.), *The Fluid Dynamics of Climate*,
CISM International Centre for Mechanical Sciences
DOI 10.1007/ 978-3-7091-1893-1_1 © CISM Udine 2016

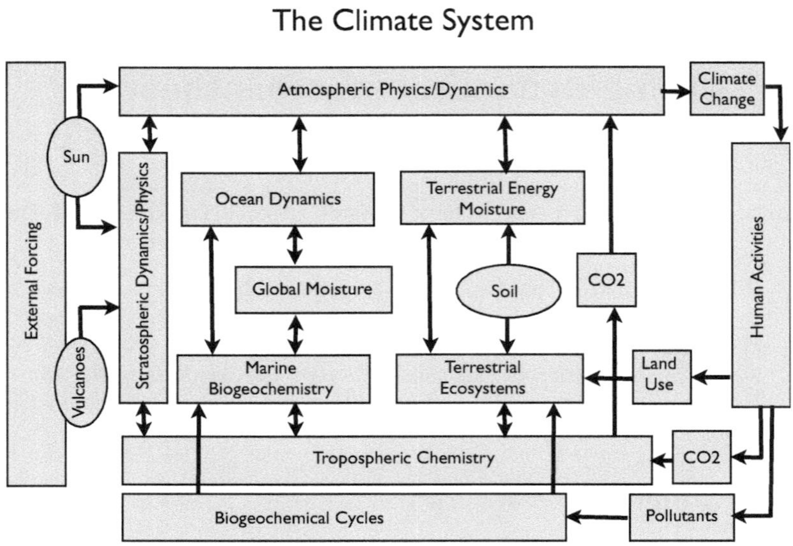

Figure 1. *A schematic of the organization of the climate system, showing the different component and their connections (simplified from Bretherton (1988)).*

seconds (e.g. formation of cloud droplets) to a few days (e.g. dissipation of midlatitude weather systems). For the ocean, these scales range from a few months (e.g. upper layer ocean) to thousands of years (e.g. deep ocean circulation adjustment). The cryosphere has an even larger range since sea ice processes are much faster than those of ice on land. The time scales of the biosphere also have a very wide range and those of the lithosphere (e.g. motion of continents) are up to millions of years.

In addition, feedbacks between the different components may also introduce new time scales of variability. One of the most important examples is the coupling between the equatorial ocean and the global atmosphere, which introduces the interannual time scale of variability associated with the El Niño/Southern Oscillation (ENSO). For feedbacks to occur, it is important to know the relevant strengths of the coupling between the different subsystems.

To determine how the Earth system develops as a whole, humans included, appears an impossible task. While for the physical components of the system, there are basic mathematical equations (i.e., conservation principles) available, such equations are (still) lacking for many other components of the system, such as for the development of ecosystems. Missing such equations on highly relevant components and facing the problem of many spatial and temporal scales of the system, we necessarily must turn to more modest approaches.

The issue of trying to determine the development of the Earth system is closely related to the question one asks. Is the question related to the development of the El Niño over the next few months, is it to predict when the next ice age will occur, or is it about the changes of the temperature in Western Europe over the next 50 years due to the increase in atmospheric greenhouse gases? Once the question has been fixed, we can immediately take advantage of the fact that processes can be differentiated according to the spatial and temporal scales and hence one can make adequate approximations. For example, we can make a very good approximation by assuming that the continents are fixed when studying the development of the weather over the next few days.

1.1 Stochastic Dynamical Systems

Let us focus on a phenomenon with a characteristic time scale T occurring predominantly over a spatial scale L (Fig. 2). As an example, we can think of the interannual variability associated with the present-day El Niño variability with spatial patterns extending over a large part of the Pacific basin, i.e., $T \approx 5$ years and $L \approx 10^7$ km. In a stochastic dynamical systems view of this phenomenon, all processes on time scales $\tau \gg T$ can be assumed to be fixed in time. For example, for present-day El Niño prediction, the ocean bathymetry can be fixed to present-day values and orbital variations in insolation can be neglected.

How do we handle the processes occurring on much smaller time and spatial scales, such as wind waves on the surface of the Pacific Ocean? This is in general a tricky issue as collective behavior due to small-scale processes certainly can influence the large-scale behavior. There is no general theory to cope with the small-scale processes apart from explicitly modeling them. When this is impossible one can resort to several options, usually referred to as parameterization of the small scales into large-scale descriptions. Parameterizations may be deterministic (i.e., given by explicit relationship between variables) or they may be stochastic. In the latter case, a stochastic model of the small scales has to be proposed. Both this stochastic model and the deterministic parameterizations are in most cases (at least partially) based on observations.

What results from this stochastic dynamical systems framework (Fig. 2) is a set of mathematical (in general: stochastic partial differential) equations

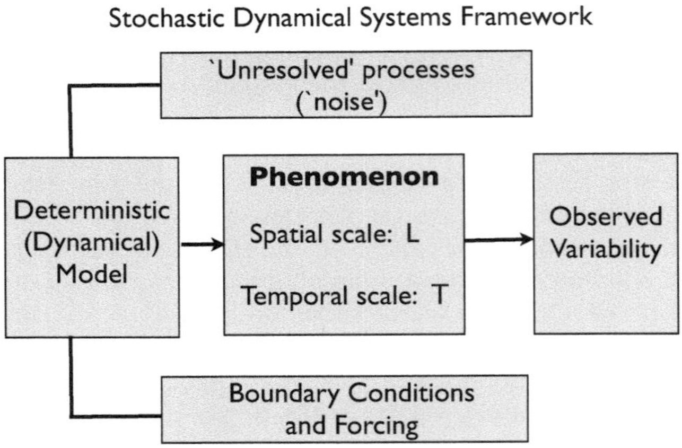

Figure 2. *A schematic of the stochastic dynamical systems framework of climate variability. A deterministic dynamical system represents the particular phenomenon on a dominant spatial scale L and a temporal scale T. Smaller and faster scale processes (labelled here as 'unresolved' by the deterministic system) are represented as 'noise', while slower processes are fixed into the boundary conditions.*

for the description of the phenomena at the scales T and L. The equations for these large scales contain parameterizations in which the behavior of the small scales are represented. Boundary conditions are formulated at areas where development of the system is much slower. Because of the cyclic nature of the insolation entering the climate system, the set of equations may contain a periodic forcing component. The response of this (in general nonlinear) periodically forced stochastic dynamical system can then be compared with data sets from the instrumental record and from proxy records.

1.2 Hierarchy of models

Climate phenomena are studied using observations and climate models. As we cannot investigate these phenomena in the laboratory, climate models are a central component of climate research. A wide range of models is in use with on one hand 'very simple' conceptual climate models and on the other hand 'very complex' state-of-the-art Global Climate Models (GCMs). It would be impossible (and also useless) to try to provide an overview of

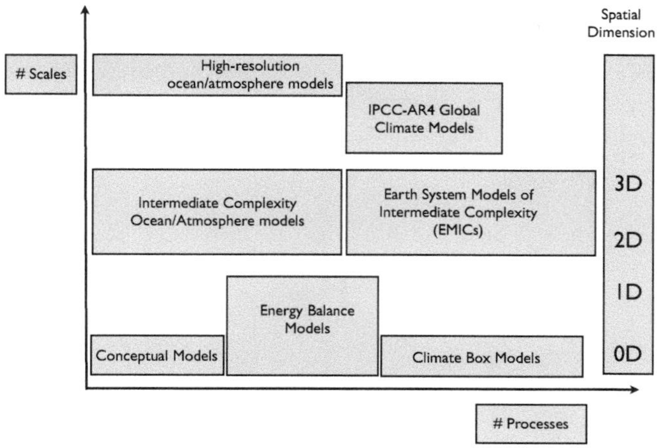

Figure 3. *'Classification' of climate models according to the two model traits: number of processes and number of scales. There is of course overlapping between the different model types, but for simplicity they are sketched here as non-overlapping. Here 0D (zero dimensional), 1D (one dimensional), 2D (two dimensional) and 3D (three dimensional) indicate the spatial dimension of the model.*

all the models around. However, general notions on the use and importance of a climate modeling hierarchy can be given.

As scales and processes are so important properties of climate phenomena, it motivates to classify climate models using these two traits (Fig. 3). Here the trait 'scales' refers to both spatial and temporal scales as there exists a relation between both: on smaller spatial scales usually faster processes take place. 'Processes' refers to either physical, chemical or biological processes taking place in the different climate compartments (atmosphere, ocean, cryosphere, biosphere, lithosphere).

Models with a limited number of processes and scales are usually referred to as conceptual climate models. In these models only very specific interactions in the climate system are described. A typical example is a box model; one of the simplest ones was used by Stommel (1961) to study the stability of the ocean's thermohaline circulation. A higher spatial resolution and inclusion of more processes will give models located in the right upper part of Fig. 3. In a GCM, the atmosphere, ocean, ice and land components are divided into grid boxes (Fig. 4). Over such a 3D grid box we consider the budgets of momentum, mass and for example heat. Momentum budgets basically follow from Navier-Stokes equations formulated for air and water. The difference on what goes into a box minus what goes out of that box leads to an increase/decrease of a particular quantity, say temperature. Once the distribution of a quantity is known at a certain time than these budgets provide an evolution equation to determine the quantity some time later.

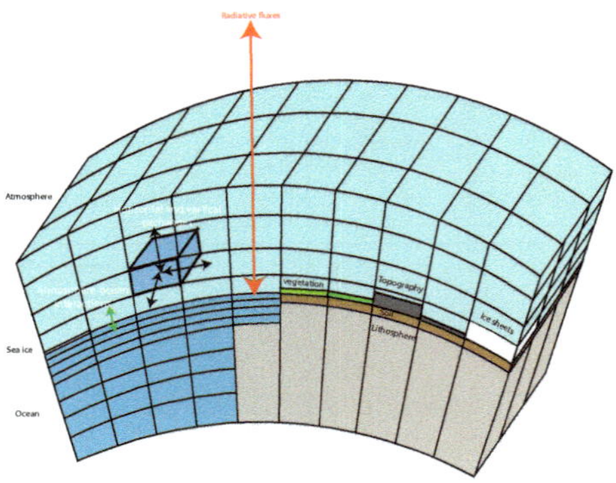

Figure 4. *A typical structure of a GCM; the number of grid boxes in each of the components determines the spatial resolution of the model (figure from http://stratus.astr.ucl.ac.be/textbook/ and courtesy of Hugues Goosse) .*

The advantage of more boxes is that we resolve the quantity better (more points in a certain area). With an increasing number of grid boxes, however, the time development of an increasing number of quantities (at each grid box) has to be calculated. The same holds for the number of processes included in a GCM: more processes simply means more calculations. Also the longer time period over which we want to compute the development of each quantity the longer it takes to do the calculation on a computer.

The state-of-the-art GCMs are located above the Earth System Models of Intermediate Complexity (EMICs) in Fig. 3 because they represent a larger number of scales (Claussen and coauthors, 2002). Compared to GCMs, the ocean and atmosphere models in EMICs are strongly reduced in the number of scales. For example, the atmospheric model may consist of a quasi-geostrophic or shallow-water model and the ocean component may be a zonally averaged model. The advantage of EMICs is therefore that they are computationally less demanding than GCMs and hence many more long-time scale processes, such as land-ice and carbon cycle processes can be included. Each of the individual component models of EMICs may also be used to study the interaction of a limited number of processes. Such models are usually referred to as Intermediate Complexity Models (ICMs). A prominent example is the Zebiak-Cane model of the El Niño/Southern Oscillation phenomenon (Zebiak and Cane, 1987). In time, the GCMs of today will be the EMICs of the future and the state-of-the-art GCMs will shift towards the upper right corner in Fig. 3.

2 ENSO variability

About once every four years, the sea surface temperature in the equatorial eastern Pacific is a few degrees higher than normal (Philander, 1990). Near the South American coast, this warming of the ocean water is usually at its maximum around Christmas. Long ago, Peruvian fishermen called it El Niño, the Spanish phrase for the "Christ Child".

2.1 Phenomena

During the last decades, El Niño has been observed in unprecedented detail thanks to the implementation of the TAO/TRITON array and the launch of satellite-borne instruments (McPhaden and coauthors, 1998). The relevant quantities to characterize the state in the equatorial ocean and atmosphere are sea level pressure, sea surface temperature (SST), sea level height, surface wind and ocean sub-surface temperature.

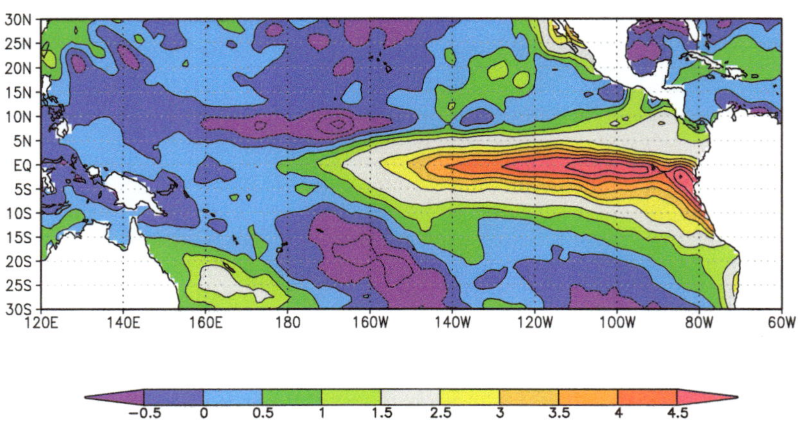

Figure 5. *Sea-surface temperature anomaly field (with respect to the 1982-2010 mean) of December, 1997, at the height of the 1997/1998 El Niño. Data from NOAA, see http://www.emc.ncep.noaa.gov/research/cmb/sst_analysis/.*

The annual mean state of the equatorial Pacific sea-surface temperature is characterized by the zonal contrast between the western Pacific "warm pool" and the "cold tongue" in the eastern Pacific. The mean temperature in the eastern Pacific is approximately 23°C, with seasonal excursions of about 3°C. What makes El Niño unique among other interesting phenomena of natural climate variability is that it has both a well-defined spatial pattern and a relatively well-defined time scale. The pattern of the sea-surface temperature anomaly for December 1997 is plotted in Fig. 5a and shows a large area where the SST is larger than average.

A common index of this sea-surface temperature anomaly pattern is the NINO3 index, defined as the sea-surface temperature anomaly averaged over the region 5°S –5°N, 150°W–90°W. In the time series (blue curve in Fig. 6a), the high NINO3 periods are known as El Niño's and the low NINO3 periods as La Niña's. There is no clear-cut distinction between El Niño's, La Niña's, and normal periods, rather the system exhibits continuous fluctuations of varying strengths and durations with an average period of about 4 years (blue curve in Fig. 6b).

Changes in the tropical atmospheric circulation are strongly connected to changes in sea-surface temperature . The red curve in Fig. 6a is the normalized pressure difference between Tahiti (Eastern Pacific region) and Darwin; this index is referred to as the Southern Oscillation Index (SOI). It measures the variations in the tropical surface winds, dominated by the trade winds. When the SOI is negative (positive) , the pressure in Tahiti is relatively low (high) with respect to that in Darwin and hence the trade winds are weakened (strengthened). The anti-correlation of the NINO3 index and the SOI is obvious from Fig. 6a and the spectrum of the SOI in Fig. 6b also shows a similar broad peak as the NINO3 index (centered at about 4 year). As El Niño and the Southern Oscillation are one phenomenon, it is referred to as the ENSO phenomenon.

There are relations between the seasonal cycle, the spatial sea-surface temperature pattern of the annual-mean state and the El Niño variability. Large sea-surface temperature anomalies often occur within the cold tongue region. In addition, El Niño is to some extent phase-locked to the seasonal cycle as most El Niño's and La Niña's peak around December (Fig. 7). The root mean square of the NINO3 index is almost twice as large in December than in April.

When one considers the spectrum of the NINO3 index (Fig. 6b), also energy is found at lower frequencies, in particular in the decadal-to-interdecadal range (Jiang et al., 1995; Zhang et al., 1997; Fedorov and Philander, 2000). The strength of El Niño before the mid-1970s appears to be smaller than that after this period; this transition is sometimes referred to as the 'Pacific

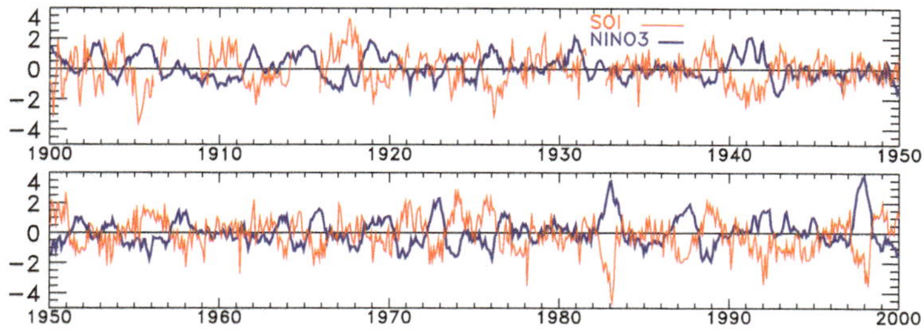

(a)

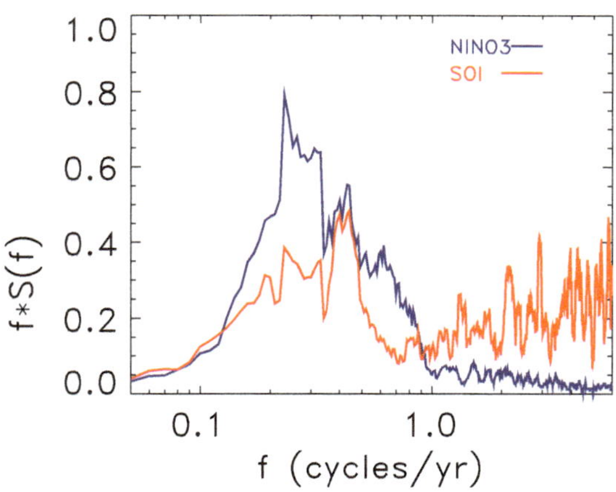

(b)

Figure 6. *(a) Time series of the NINO3 index (dark) and SOI (light) over the years 1900-2000. (b) Spectrum of the NINO3 index (dark) and SOI (light) in (a), where on the vertical axis the product of frequency f and the spectral power $S(f)$ is plotted (figure based on Dijkstra and Burgers (2002)).*

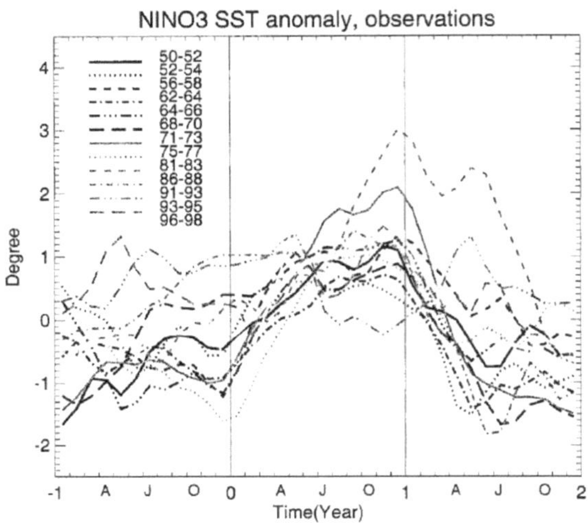

Figure 7. *NINO3 index for 13 observed ENSO events from 1950 to 1998 with the mean seasonal cycle of this period removed (from Neelin et al. (2000)). The curves are aligned based on the year of the peak warm phase (at year 1).*

climate shift' (Trenberth, 1997). According to NCEP data, the standard deviation of the SOI (NINO3) for 1951-1975 is 1.64 (0.81), to be compared for 1976-2000 where it is 1.84 (1.00). The spatial pattern of these (multi)decadal changes is fairly similar to that of the interannual variability, but the sea-surface temperature anomalies at the eastern side of the basin extend from the equator to midlatitudes (Zhang et al., 1997).

2.2 A Minimal Model

One of the first models that was able to reasonably simulate ENSO was that of Zebiak and Cane (1987). In its original version, an annual mean state or seasonal cycle of both ocean and atmosphere was obtained from observations and within the model the evolution of anomalies with respect to this reference state were computed. We will refer below to this Intermediate Complexity Model (ICM) as the ZC model.

The model captures the evolution of large scale motions in the tropical ocean and atmosphere in a domain of infinite extent in the meridional direction. The ocean is bounded by meridional walls at the west ($x = 0$) and east ($x = L$) coast. The ocean component of the model consists of a well-mixed layer of mean depth H_1 embedded in a shallow-water layer of mean depth $H = H_1 + H_2$ having a constant density ρ (Fig. 8). The lower boundary of the shallow-water layer corresponds to the region of a sharp vertical gradient in temperature and is referred to as the thermocline. Only long wave motions above the thermocline are considered and the deep ocean (having a constant density $\rho + \Delta\rho$) is assumed to be at rest.

We start with the ocean model component. In a first step, the total horizontal velocity field is split into a mean $\mathbf{u}$ over the mixed layer and shallow-water layer and a difference velocity, the surface velocity $\mathbf{u}_s$. The mean velocity $\mathbf{u} = (u, v)$ satisfies a reduced gravity model on the equatorial β plane (with planetary vorticity gradient β_0). With the surface wind stress field given by (τ^x, τ^y), these equations are given by

$$\frac{\partial u}{\partial t} + a_m u - \beta_0 y v + \frac{\partial h}{\partial x} = \frac{\tau^x}{\rho H}, \tag{1a}$$

$$\beta_0 y u + g' \frac{\partial h}{\partial y} = \frac{\tau^y}{\rho H}, \tag{1b}$$

$$\frac{\partial h}{\partial t} + a_m h + c_o^2 (\frac{\partial u}{\partial x} + \frac{\partial v}{\partial y}) = 0, \tag{1c}$$

where h is the total upper layer thickness. In addition, we have used a linear damping coefficient a_m and the reduced gravity $g' = g\Delta\rho/\rho$. The quantity $c_0 = \sqrt{g'H}$ is the equatorial Kelvin wave speed. In the long-wave approximation, the boundary conditions are

$$\int_{-\infty}^{\infty} u(0, y, t) dy = 0 \ , \ u(L, y, t) = 0. \tag{2}$$

The equations for the surface layer velocities $\mathbf{u}_s$ are the Ekman-layer balances

$$a_s u_s - \beta_0 y v_s = \frac{H_2}{H} \frac{\tau^x}{\rho H_1}, \tag{3a}$$

$$a_s v_s + \beta_0 y u_s = \frac{H_2}{H} \frac{\tau^y}{\rho H_1}, \tag{3b}$$

where a_s is again a linear damping coefficient. With $u_1 = u_s + u$ and $v_1 = v_s + v$ being the velocity in the mixed layer, the vertical velocity component is determined from continuity as

$$w_1 = H_1 (\frac{\partial u_1}{\partial x} + \frac{\partial v_1}{\partial y}). \tag{4}$$

The evolution of the mixed layer temperature T is governed by the equation

$$\frac{\partial T}{\partial t} + a_T(T - T_0) + \frac{w_1}{H_u}\mathcal{H}(w_1)(T - T_s(h)) + u_1\frac{\partial T}{\partial x} + v_1\frac{\partial T}{\partial y} = 0, \tag{5}$$

where $\mathcal{H}$ is a continuous approximation of the Heaviside function. The second term in (5) is usually referred to as the Newtonian cooling term, with inverse damping time scale a_T, representing all processes as horizontal mixing, sensible and latent heat surface fluxes, and long wave and shortwave radiation. T_0 is the temperature of radiative equilibrium which is realized in the absence of large-scale horizontal motion in the upper ocean and atmosphere. The third term in (5) models the heat flux due to upwelling through the total velocity w_1 and the approximate vertical temperature gradient $(T - T_s(h))/H_u$. The subsurface temperature (T_s) depends on the thermocline deviations and models the effect that heat is transported upwards (if $w_1 > 0$) when the cold water is further from the surface. An explicit expression, which is often used, is

$$T_s(h) = T_{s0} + (T_0 - T_{s0}) \tanh\left[\frac{h + h_0}{h_1}\right], \tag{6}$$

where h_0 is an offset value, T_{s0} is the subsurface temperature for $h = -h_0$ and h_1 controls the steepness of the transition as h passes through $-h_0$. The last two terms in (5) represent horizontal advection of heat.

Sea surface temperature anomalies cause wind anomalies and one of the simplest models to represent the wind response can be determined by the Gill (1980) model. In this description, the atmospheric zonal and meridional boundary-layer velocities (U, V) and geopotential Θ satisfy

$$\frac{\partial U}{\partial t} + a_M U - \beta_0 y V - \frac{\partial \Theta}{\partial x} = 0, \tag{7a}$$

$$\frac{\partial V}{\partial t} + a_M V + \beta_0 y U - \frac{\partial \Theta}{\partial y} = 0, \tag{7b}$$

$$\frac{\partial \Theta}{\partial t} + a_M \Theta - c_a^2\left(\frac{\partial U}{\partial x} + \frac{\partial V}{\partial y}\right) = \alpha_T(T - T_r), \tag{7c}$$

where a_M is a damping coefficient and c_a^2 is the atmospheric equatorial Kelvin wave speed. The right hand side of (7c) is proportional to the ocean-atmosphere heat flux Q_{oa} and T_r is a reference temperature.

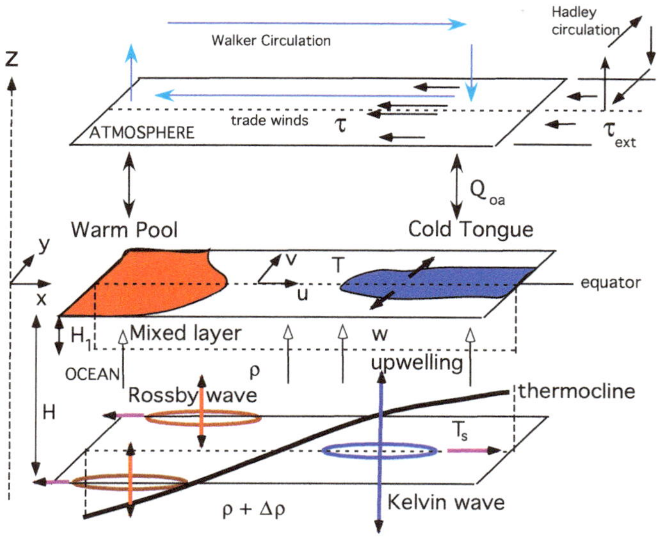

Figure 8. *Overview of the oceanic and atmospheric processes of the equatorial coupled ocean-atmosphere system which are represented in the ZC model.*

Atmospheric velocity anomalies give wind stress anomalies to the ocean surface according to

$$\tau^x = \gamma U \ ; \ \tau^y = \gamma V, \tag{8}$$

which completes the coupling between the atmosphere and ocean. In Zebiak and Cane (1987), many more details of the model set-up are provided.

Ocean processes The effects of the wind stress on the upper ocean are threefold. As described above, the dominantly easterly wind stress causes water to pile up near the western part of the basin. This induces a higher pressure in the upper layer western Pacific than that in the eastern Pacific and consequently a shallowing of the thermocline towards the east. Second, the winds cause divergences and convergences of mass in the upper ocean, due to the frictional (Ekman) boundary layer in which the momentum input is transferred. North of the equator, the trade winds cause an Ekman transport to the right of the wind away from the equator. Similarly, south of the equator the Ekman mass transport is away from the equator. With a wind stress amplitude of 0.1 Pa, a typical value of the vertical velocity is a few meters per day. Finally, the wind stress is responsible for the presence of the upper ocean currents.

When the amplitude and/or the direction of the wind stress changes, the upper ocean adjusts through wave dynamics. The most important waves involved in this adjustment process are:

(i) Equatorial Kelvin waves. For such a wave, the meridional structure of the thermocline is maximal at the equator. Its amplitude decays exponentially in meridional direction with a decay scale λ_0 of about 300 km. The zonal velocity has the same spatial structure as the thermocline and the meridional velocity is zero. The group/phase velocity of these non-dispersive waves, $c_o \approx 2$ ms^{-1}. It takes about $\tau_K = 3$ months to cross the Pacific basin of width $L = 15,000$ km.

(ii) Long equatorial Rossby waves. For such a wave, the meridional thermocline structure has an off-equatorial maximum. For each of these waves, the meridional velocity is much smaller than the zonal velocity. The group/phase velocity of these non-dispersive waves c_j, with a meridional spatial structure coupled to the index j, is $c_j = -c_o/(2j+1)$. The $j = 1$ long Rossby wave travels westward with a velocity which is 1/3 of that of the Kelvin wave and hence takes about $\tau_R = 9$ months to cross the Pacific basin.

When a Kelvin wave meets the east coast, the dominant contribution of the reflected signal are long Rossby waves. Similarly, when a Rossby wave meets the west coast, it is reflected dominantly as a Kelvin wave.

The small amplitude response of the flow in an equatorial ocean basin to a certain wind field can be described as a directly forced response and a sum of free waves to satisfy the boundary conditions (Moore, 1968; Cane and Sarachik, 1977). In the response to a time-periodic wind stress, the ocean does not only react to the instantaneous wind pattern but also to previous winds through propagation of waves. Along the equator, the ocean adjustment is mainly accomplished by relatively fast eastward Kelvin waves.

Off-equatorial adjustment is accomplished by slower Rossby waves that travel westward. The off-equatorial thermocline pattern consists partly of free Rossby waves which are adjusting to the wind and partly of a forced response which is in quasi-steady balance with the instantaneous wind stress. It is the departure of this quasi-steady balance, which is crucial to further evolution of the thermocline and provides the ocean with a memory. A measure of this memory was considered in Neelin et al. (1998) to be the difference between the actual response and that which would be, at every time, in steady balance with the instantaneous wind stress. For the wind patterns over the Tropical Pacific, this 'ocean memory' is largest near the western boundary.

Coupled feedbacks If the underlying sea surface is warm, the air above it is heated and rises. When one assumes that convection mostly occurs over the warmest water and that the adiabatic heating is dominated by heat which is released during precipitation, latent heat anomalies are proportional to sea-surface temperature anomalies, represented by the proportionality factor α_T in (7). In this case, a low level zonal wind response is found with westerly (easterly) winds to the west (east) of positive (negative) sea-surface temperature anomalies (Fig. 9). As the zonal wind stress anomalies are proportional to the zonal wind anomalies, with proportionality factor γ (8), the nature of the coupling between ocean and atmosphere can now be explained.

A sea-surface temperature anomaly will give, through local heating, a lower-level wind anomaly. The resulting wind stress anomaly on the ocean-atmosphere surface will (i) change the thermocline slope through horizontal pressure differences in the upper ocean, (ii) change the strength of the upwelling through the Ekman divergences in the upper layer and (iii) affect the upper ocean currents (u, v) in the mixed layer. The changes in velocity field and thermocline field will affect the sea surface temperature, according to (5).

The strength of the coupling between the ocean and atmosphere is determined by the combined effects of the amplitude of the zonal wind anomaly which generated by sea-surface temperature anomalies and how much of the momentum of this wind is transferred as stress to the upper ocean layer. The strength of the coupling is measured by the parameter μ which is a (dimensionless) product of α_T and γ, i.e.

$$\mu = \frac{\gamma \alpha_T \Delta T L^2}{c_o^2 c_a^2} \tag{9}$$

where the quantity ΔT is a typical zonal temperature difference over the basin and L is the zonal length of the basin.

The main positive feedbacks identified in this coupled ocean-atmosphere system are referred to the thermocline, upwelling and zonal advection feedback (Neelin, 1991). They are best illustrated by looking at the growth of very small perturbations (quantities with a tilde) on a background state (quantities with a bar). The linearized temperature equation, describing the evolution of the temperature perturbations $\tilde{T}$ can be written as

$$\begin{aligned} \frac{\partial \tilde{T}}{\partial t} + a_T \tilde{T} + \frac{\bar{w}_1}{H_u} \mathcal{H}(\bar{w}_1)(\tilde{T} - \tilde{T}_s) \quad &+ \quad \frac{\tilde{w}_1}{H_u} \mathcal{H}(\bar{w}_1)(\bar{T} - \bar{T}_s) + \\ \bar{u}_1 \frac{\partial \tilde{T}}{\partial x} + \bar{v}_1 \frac{\partial \tilde{T}}{\partial y} \quad &+ \quad \tilde{u}_1 \frac{\partial \bar{T}}{\partial x} + \tilde{v}_1 \frac{\partial \bar{T}}{\partial y} = 0, \end{aligned} \tag{10}$$

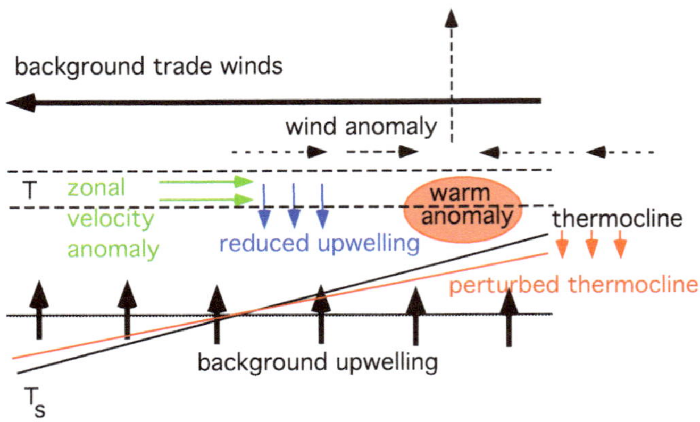

Figure 9. *Sketch to illustrate the thermocline feedback (red arrows), the upwelling feedback (blue arrows) and the zonal advection feedback (green arrows).*

The thermocline feedback is best explained by looking at a sloping thermocline in a constant upwelling ($\bar{w}_1 > 0$) ocean as sketched in Fig. 9. The sloping thermocline and the upwelling are caused by the background easterly winds. Now assume that a positive sea-surface temperature perturbation $\tilde{T}$ is present at some location, for example in the eastern part of the basin. This leads to a perturbation in the low level zonal wind which is westerly with a maximum located west of the maximum of the sea-surface temperature anomaly. Since the background winds are weakened locally, the slope of the thermocline decreases and it becomes more flat (red arrows in Fig. 9). In this case, the colder water will be closer to the surface in the west but it will be farther down in the east. Hence, the sub-surface temperature effectively increases at the level of upwelling, giving a positive heat flux perturbation at the bottom of the mixed layer. According to (10),

$$\partial \tilde{T}/\partial t \approx -\bar{w}_1(\tilde{T} - \tilde{T}_s)/H_u,$$

and when $\tilde{T}_s - \tilde{T} > 0$, the original disturbance is amplified, as $\bar{w}_1$ is positive. As $\tilde{T}_s \sim \tilde{h}$, a deeper thermocline in the east can amplify a positive temperature anomaly. The thermocline feedback is present in a transient state, but also in an balanced (adjusted) state.

To understand the upwelling feedback, consider again a positive sea-surface temperature anomaly in the eastern part of the basin and associated changes in the wind. However, now changes in the upwelling $\tilde{w}_1$, mainly through the Ekman layer dynamics, occur. Weaker easterly winds imply less upwelling and hence less colder water enters the mixed layer (blue arrows in Fig. 9). If $\tilde{w} < 0$ and the background vertical temperature gradient is stably stratified ($\bar{T} > \bar{T}_s$), then the sea-surface temperature perturbation is amplified. The latter can be seen again from (10), i.e.

$$\partial\tilde{T}/\partial t \approx -\tilde{w}_1(\bar{T} - \bar{T}_s)/H_u.$$

The zonal advection feedback arises through zonal advection of heat. Imagine a region with a strong sea surface temperature gradient, say $\partial\bar{T}/\partial x < 0$. Such a region occurs for example at the east side of the warm pool. Suppose a positive anomaly in sea-surface temperature occurs leading to westerly wind anomalies. Consequently, the zonal surface ocean current ($\tilde{u} > 0$) is intensified (green arrows in Fig. 9) leading to amplification of the positive temperature perturbation, according to (10), i.e.

$$\partial\tilde{T}/\partial t \approx -\tilde{u}_1 \, \partial\bar{T}/\partial x.$$

Part of the mixed layer zonal velocity will be due to wave dynamics and part due to Ekman dynamics.

2.3 The ENSO mode

In many early model studies on ENSO variability, the annual-mean state was simply prescribed without any consideration of the processes causing this state. For example, a flat thermocline was chosen, with a spatially constant zonal temperature gradient and no-flow in both ocean and atmosphere (Hirst, 1986; Neelin, 1991). Alternatively, an annual-mean state or a seasonal cycle derived from observations was prescribed (Zebiak and Cane, 1987).

Stability of the Annual Mean State To understand the spatial pattern of the warm pool/cold tongue, it was recognized that part of the (annual mean) wind stress in the Pacific basin is in fact related to its zonal temperature gradient (Neelin and Dijkstra, 1995). Only a small part of the wind stress, say $\vec{\tau}_{ext}$, is determined externally. The external zonal wind stress τ^x_{ext} can be taken constant, imagined as being caused by the zonally symmetric atmospheric circulation.

At zero coupling strength ($\mu = 0$), the ocean circulation and consequently the sea-surface temperature is determined by the external zonal wind stress τ^x_{ext}. A small easterly wind stress causes a small amount of upwelling and a small slope in the thermocline. Within the ZC model variant used in Van der Vaart et al. (2000), the equatorial sea-surface temperature increases monotonically from about 26°C (Fig.10a) in the east to about

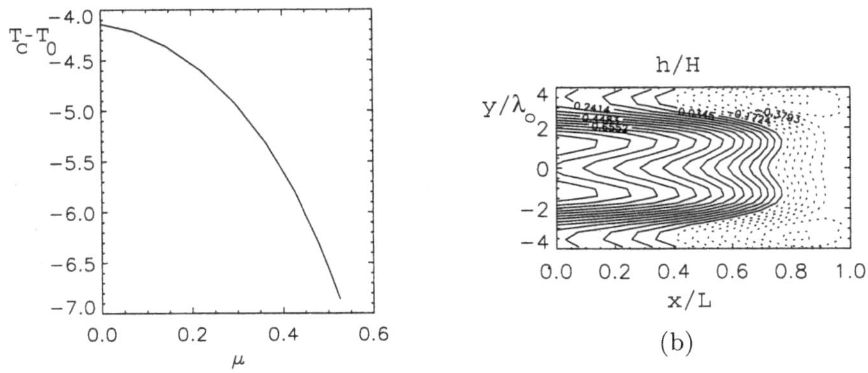

(a)

Figure 10. *(a) Eastern Pacific ($x/L = 0.8$) equatorial sea-surface temperature T_C deviation from $T_0 = 30°C$ as a function of the coupling strength μ. (b) Pattern of the thermocline depth anomaly at $\mu = 0.5$. The quantities $\lambda_0 = \sqrt{c_0/\beta_0}$ (~ 300 km) and $H = 100$ m are typical meridional and vertical scales of the thermocline (figure from Van der Vaart et al. (2000)).*

29°C in the west, in response to $\tau^x_{ext} = 0.01$ Pa. The thermocline is approximately linear at the equator, its depth is increasing westwards and it has slight off-equatorial maxima.

Additional easterly wind stress occurs due to coupling because of the zonal temperature gradient. In Neelin and Dijkstra (1995), the zonal wind stress was represented as

$$\tau^x = \tau^x_{ext} + \mu \mathcal{A}(T - T_0), \qquad (11)$$

where $\mathcal{A}$ is a short representation of the atmosphere model operator and T_0 is again the radiative equilibrium temperature. Increasing μ leads to larger upwelling and a larger thermocline slope, strengthening the cold tongue in the eastern part of the basin. The temperature of the cold tongue T_C is shown in Fig. 10a as a function of μ. The zonal scale of the cold tongue is set by a delicate balance of the different coupled feedbacks. The thermocline field at $\mu = 0.5$ (Fig. 10b) displays the off-equatorial maxima and a deeper (shallower) equatorial thermocline in the west (east). This shows that coupled processes are involved in the annual mean spatial patterns of the Pacific climate system (Dijkstra and Neelin, 1995).

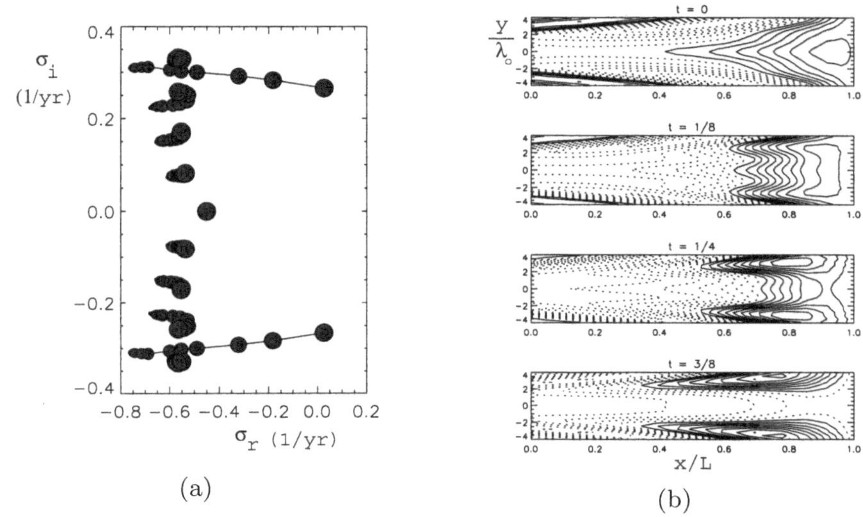

Figure 11. *(a) Plot of the eigenvalues for the six leading eigenmodes in the $(Re(\sigma), Im(\sigma))$-plane. Values of the coupling strength μ are represented by dot size (smallest dot is the uncoupled case ($\mu = 0$) for each mode, largest dot is the fully coupled case at the stability boundary ($\mu_c = 0.5$), the location of the Hopf bifurcation). (b) Planforms of the thermocline depth anomaly at several phases of the 3.7 year oscillation; time t=1/2 refers to half a period. Drawn (dotted) lines represent positive (negative) anomalies (figure from Van der Vaart et al. (2000)).*

Spatial pattern of the ENSO mode Taking the annual-mean state as a background state, we are now interested in its sensitivity to small perturbations. Necessary conditions for instability can be obtained by determining the linear stability boundary through normal mode analysis. In such an analysis, an arbitrary perturbation is decomposed in modes (e.g., Fourier modes) and the growth or decay of each of these modes is investigated. When the background state is stationary, the time-dependence of each mode is of the form $e^{\sigma t}$, where $\sigma = \sigma_r + i\sigma_i$ is the complex growth rate. With μ as the control parameter, then the linear stability boundary μ_c is the first value of μ where $\sigma_r = 0$ for one particular normal mode. A mode with $\sigma_i \neq 0$ is oscillatory (with a period $2\pi/\sigma_i$) while a mode with $\sigma_i = 0$ is called stationary. A Hopf bifurcation occurs when the growth rate of an oscillatory mode changes sign. If the growth factor of a mode is positive, sustained oscillations will occur. If the growth factor of one of these modes is negative, it will be damped.

In early studies (Hirst, 1986), the stability of a prescribed spatially constant mean state in a periodic ocean basin was considered. In this case, both uncoupled ocean and atmospheric Rossby and Kelvin waves may destabilize due to the coupled processes. The resulting unstable modes are travelling waves, which have an interannual oscillation period for wavelengths which are in the order of the basin size (Philander et al., 1984). Also a slow westward propagating mode, a so-called SST-mode, may become unstable. This mode is not related to wave dynamics in either the atmosphere or the ocean, but to adjustment processes of SST (Neelin, 1991).

The linear stability of prescribed zonally varying annual-mean states within a bounded basin was investigated in Jin and Neelin (1993). In the uncoupled case, two distinct sets of modes appear. One set is primarily related to the time scales of sea-surface temperature change (the above mentioned SST-modes) and the other set is related to time scales of ocean adjustment (ocean-dynamics modes). Depending on the other parameters in the model, the mean state can become unstable to stationary instabilities as well as oscillatory ones. In the parameter regime considered 'most realistic', referred to as the 'standing oscillatory' regime, merging of an oscillatory ocean dynamics mode with a stationary SST-mode occurs. This leads to slightly growing (so-called mixed SST/ocean-dynamics) modes, which inherit their spatial structure from the stationary SST-mode, but for which their interannual oscillation period is set by ocean sub-surface dynamics (Neelin et al., 1994).

The linear stability of the fully coupled annual-mean states in Fig. 10 was studied in Van der Vaart et al. (2000). In Fig. 11a, the path of six modes – which become leading eigenmodes at high coupling – is plotted as a function of the coupling strength μ. A larger dot size indicates a larger value of μ and both oscillation frequency σ_i and growth rate σ_r of the modes are given in year^{-1}.

The growth rate of one oscillatory mode becomes positive as μ is increased beyond $\mu_c \approx 0.5$, which is the location of the first Hopf bifurcation. The equatorial sea-surface temperature pattern of this mode displays a nearly standing oscillation for which the spatial scale is confined to the cold tongue of the mean state. The wind response is much broader zonally and is in phase with the sea-surface temperature anomaly. In the spatial structure of the thermocline field (plotted in Fig. 11b at several phases of the oscillation) the eastward propagation of equatorial anomalies, their reflection at the eastern boundary and subsequent off-equatorial westward propagation can be distinguished.

Using prescribed annual-mean states, Fedorov and Philander (2000) provide an overview of the dependence of the growth factor (Fig. 12a) and

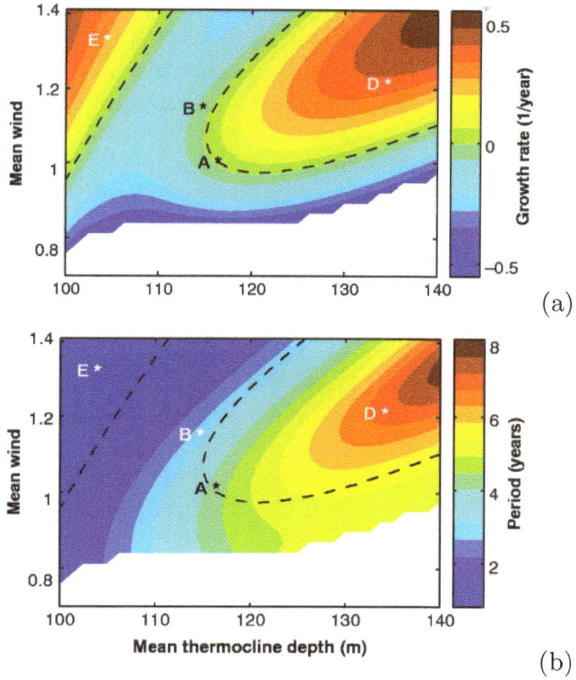

Figure 12. *Dependence of (a) growth rate (σ_r in yr^{-1}) and (b) period ($2\pi/\sigma_i$ in yr^{-1}) of the ENSO mode on the mean zonal wind stress (in units of 0.5 $cm^2 s^{-2}$) and the mean thermocline depth (in m) (figure from Fedorov and Philander (2000)).*

period (Fig. 12b) of the ENSO mode on the background conditions. The dashed curve in Fig. 12 represent zero growth (neutral conditions, $\sigma_r = 0$) of the ENSO mode, which is just a path of Hopf bifurcations in the two-parameter plane. Present-day estimates for the Pacific background state correspond to the area near the points A and B, with a period of 2-6 years and near neutral conditions.

2.4 Mechanisms of ENSO variability

Conceptual models are available to explain the physics of the oscillation period and the growth of the ENSO mode. Most of these share the ideas that one of the positive feedbacks acts to amplify sea-surface temperature anomalies and that adjustment processes in the ocean eventually cause a delayed negative feedback. These common elements are grouped together in the so-called delayed oscillator mechanism. The differences between the more detailed views are subtle (Jin, 1997) and are related to the role of the boundary wave reflections, the importance of the adjustment processes of sea-surface temperature, the dominant feedback which is responsible for amplification of anomalies and the view of the dynamical adjustment processes in the ocean. There are four type of ENSO "oscillators", the (i) 'classical' delayed oscillator, (ii) the recharge oscillator, (iii) the western Pacific oscillator and (iv) the advective/reflective oscillator. Below, we only describe (i) and (ii); the other two oscillators are described elsewhere (Wang, 2001; Wang and Picaut, 2004).

The classical delayed oscillator In this view, the eastern boundary reflection is unimportant, the thermocline feedback is dominant and individual Kelvin and Rossby waves control ocean adjustment. A minimal model (Suarez and Schopf, 1988; Battisti and Hirst, 1989; Munnich et al., 1991) representing this behavior is a differential delay equation of the form

$$\frac{dT(t)}{dt} = a_1 h_e(x_c, t - \tau_1) - a_2 h_o(x_c, t - \tau_2) - a_3 T^3(t). \tag{12}$$

In this equation, the $a_i, i = 1, 2, 3$ are constants, T is the eastern Pacific temperature anomaly, which is influenced by midbasin (at $x = x_c$) equatorial thermocline anomalies h_e and off-equatorial anomalies h_o. Furthermore, $\tau_1 = \frac{\tau_K}{2}$ and $\tau_2 = \frac{\tau_R}{2} + \tau_K$ where τ_K and τ_R are the basin crossing times of the Kelvin wave and the gravest ($j = 1$) Rossby wave.

When the equatorial Kelvin wave, which deepens the thermocline, arrives in the eastern Pacific, local amplification of temperature perturbations occurs through the thermocline feedback, represented by the first term in the right hand side of (12). The wind response excites (off-equatorial) Rossby waves which travel westwards, reflect at the western boundary and return as a Kelvin wave which rises the thermocline and provides a delayed negative feedback, represented by the second term in (12). The delay τ_2 is the time taken for the Rossby wave to travel from the center of wind response to the western boundary plus the time it takes the reflected Kelvin wave to cross the basin. The nonlinear term in (12) is needed for equilibration of the temperature anomaly to finite amplitude.

The recharge oscillator view In this theory, the ocean adjustment is viewed as being caused by a collective of Kelvin and Rossby waves and adjustment of sea-surface temperature through surface layer processes is also important. Consider again a positive sea-surface temperature anomaly in the eastern part of the basin which induces a westerly wind response. Through ocean adjustment, the slope in the thermocline is changed giving a deeper eastern thermocline. Hence, through the thermocline feedback, the sea-surface temperature anomaly is amplified which brings the oscillation to the extreme warm phase (Fig. 13a). Because of ocean adjustment, a nonzero divergence of the zonally integrated mass transport occurs and part of the equatorial heat content is moved to off-equatorial regions. This exchange causes the equatorial thermocline to flatten and reduces the eastern temperature anomaly (again the thermocline feedback plays a role) and consequently the wind stress anomaly vanishes (Fig. 13b).

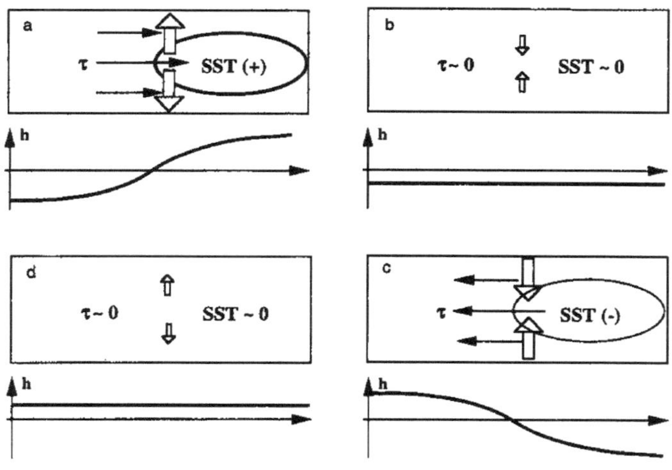

Figure 13. *Sketch of the different stages of the recharge oscillator (Jin, 1997).*

Eventually a non-zero negative thermocline anomaly is generated, which allows cold water to get into the surface layer by the background upwelling. This causes a negative sea-surface temperature anomaly leading through amplification to the cold phase of the cycle (Fig. 13c). Through adjustment, the equatorial heat content is recharged (again the zonally integrated mass transport is nonzero) and leads to a transition phase with a positive zonally integrated equatorial thermocline anomaly (Fig. 13d). This recharge oscillator view can be easily adapted to include the zonal advection feedback (Jin and An, 1999) as the zonal advection feedback and the thermocline feedback induce the same tendencies of sea-surface temperature anomalies.

3 The Atlantic Multidecadal Oscillation

After the excursion to the equatorial Pacific in the previous section, we now consider climate variability on multidecadal time scales in the North Atlantic.

3.1 Basic phenomena

The first analyses of multidecadal variability in the North Atlantic (Schlesinger and Ramankutty, 1994; Kushnir, 1994) were based on observed sea-surface temperature (SST) and indicated the existence of variability on a time scale of 50-70 years. This variability was named the Atlantic Multidecadal Oscillation (AMO) by Kerr (2000). An AMO index, defined by Enfield et al. (2001) as the ten-year running mean of detrended Atlantic SST anomalies north of the equator, is plotted in Fig. 14. Warm periods were in the 1940s and from 1995 up to the present, whereas during the 1970s the North Atlantic was relatively cold.

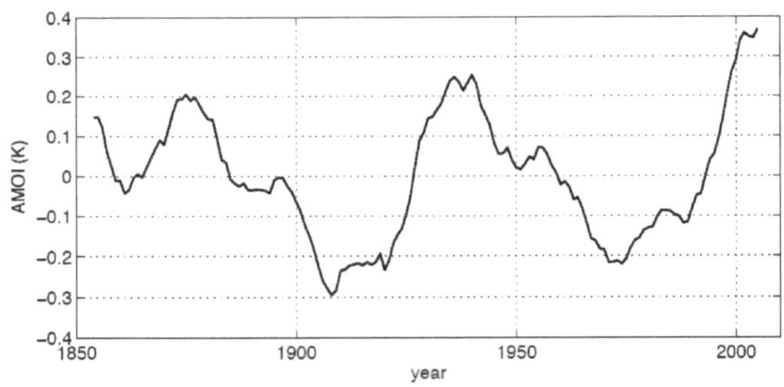

Figure 14. *The AMO index (the ten-year running mean of detrended Atlantic sea surface temperature anomalies north of the equator) of the instrumental record (using data from the HadSST2 dataset).*

Low-frequency variability in the North Atlantic SST has been determined from proxy data stretching back at least 300 years (Delworth and Mann, 2000) and within this data there is a statistically significant peak above a red-noise background at about 50 years. From recent Greenland ice-core analysis, where five overlapping records between the years 1303 and 1961 are available with annual resolution, significant multidecadal peaks in the spectrum were found (Chylek et al., 2011).

A first impression of the pattern of the AMO was obtained from an analysis of SSTs in the North Atlantic from the instrumental record (Kushnir, 1994). Fig. 15 shows the difference between the average SST during the relatively warm years 1950-1964 and the relatively cool years 1970-1984. There is a negative SST anomaly near the coast of Newfoundland and a positive SST anomaly over the rest of the North Atlantic basin. Because

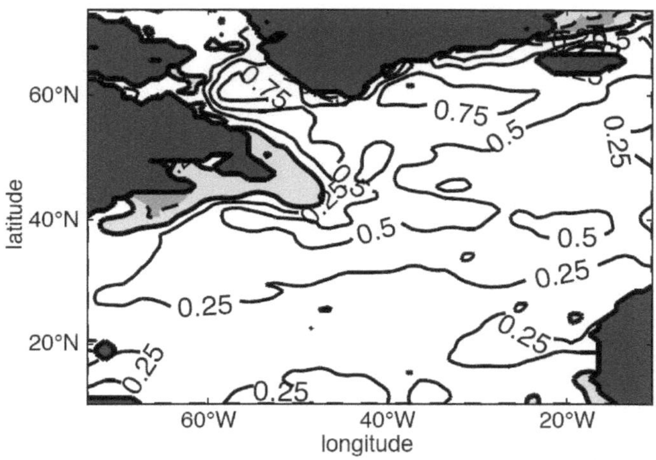

Figure 15. *An impression of the AMO pattern as shown by the difference in observed average North Atlantic SST between the periods 1950-1964 (warm period) and 1970-1984 (cold period). Units are in °C and negative values are shaded.*

of the relatively short observational time series of the instrumental record (~150 year of SST and SLP), it is difficult to extract a dominant pattern of multidecadal variability with much confidence. Kaplan et al. (1997) and Delworth and Greatbatch (2000) present a reconstruction of a signal with a ~ 50 year period which shows a near standing pattern in SST and SLP. The SST pattern is basin wide with the largest anomalies appearing south of Greenland.

A persistent, large-scale temperature anomaly over an ocean basin the size of the North Atlantic represents a significant amount of heat. It comes as no surprise, therefore, that the AMO has an effect on the climate of the surrounding land masses. Sutton and Hodson (2005) found that sea level pressure, precipitation and temperature over Europe and North America, particularly during June-August, are affected by the AMO. The AMO has also been linked to rainfall and thence to river flows in the USA (Enfield et al., 2001). There is a negative correlation between the AMO index and rainfall over the Mississippi basin and a positive correlation between the AMO index and rainfall in Florida. This means that there is on average less (more) rain over the Mississippi catchment area and more (less) rain over Florida when the AMO index is high (low). Positive correlations have also been found between the AMO index and rainfall in the Sahel and between the AMO index and the strength of the Indian summer monsoon (Zhang and Delworth, 2006; Knight et al., 2006; Feng and Hu, 2008).

A basic theory of AMO should explain (i) the processes leading to the shape and amplitude of the spatial pattern of the SST anomalies, and (ii) the physics of the dominant time scale of variability.

3.2 Minimal model

Based on results from CGCM studies (Delworth et al., 1993), a minimal model of the AMO was formulated by Greatbatch and Zhang (1995) and Chen and Ghil (1996). It consists of a flow in an idealized three-dimensional northern hemispheric sector basin forced only by a prescribed heat flux. It turns out that these flows are susceptible to large-scale multidecadal time scale oscillatory instabilities (Huck et al., 1999; Huck and Vallis, 2001; Huck et al., 2001; Kravtsov and Ghil, 2004). The mechanism of these oscillations and the associated spatial SST patterns is presented in this section.

We consider ocean flows in a model domain on the sphere bounded by the longitudes $\phi_w = 286°$E (74°W) and $\phi_e = 350°$E (10°W) and by the latitudes $\theta_s = 10°$N and $\theta_n = 74°$N; the ocean basin has a constant depth H. The flows in this domain are forced by a restoring heat flux Q_{rest} (in Wm^{-2}) given by

$$Q_{rest} = -\lambda_T(T^* - T_S), \tag{13}$$

where λ_T (in Wm^{-2}K^{-1}) is a constant surface heat exchange coefficient. The heat flux Q_{rest} is proportional to the temperature difference between the ocean temperature T^* taken at the surface and a prescribed atmospheric temperature T_S, chosen as

$$T_S(\phi, \theta) = T_0 + \frac{\Delta T}{2} \cos\left(\pi \frac{\theta - \theta_s}{\theta_n - \theta_s}\right), \tag{14}$$

where $T_0 = 15°$C is a reference temperature and ΔT is the temperature difference between the southern and northern latitude of the domain. The thermal forcing is distributed as a body forcing over the first (upper) layer of the ocean having a depth H_m.

Temperature differences in the ocean cause density differences according to

$$\rho = \rho_0(1 - \alpha_T(T^* - T_0)), \tag{15}$$

where α_T is the volumetric expansion coefficient and ρ_0 is a reference density. Inertia is neglected in the momentum equations because of the small Rossby number, we use the Boussinesq and hydrostatic approximations and represent horizontal and vertical mixing of momentum and heat by constant eddy coefficients. With r_0 and Ω being the radius and angular velocity of the Earth, the governing equations for the zonal, meridional and vertical velocity u, v and w, the dynamic pressure p (the hydrostatic part has been subtracted) and the temperature $T = T^* - T_0$ become

$$
\begin{aligned}
-2\Omega\, v \sin\theta + \frac{1}{\rho_0 r_0 \cos\theta}\frac{\partial p}{\partial \phi} - A_H L_u(u,v) - A_V \frac{\partial^2 u}{\partial z^2} &= 0, && \text{(16a)}\\
2\Omega\, u \sin\theta + \frac{1}{\rho_0\, r_0}\frac{\partial p}{\partial \theta} - A_H\, L_v(u,v) - A_V \frac{\partial^2 v}{\partial z^2} &= 0, && \text{(16b)}\\
\frac{\partial p}{\partial z} - \rho_0 g \alpha_T T &= 0, && \text{(16c)}\\
\frac{1}{r_0 \cos\theta}\left(\frac{\partial u}{\partial \phi} + \frac{\partial (v\cos\theta)}{\partial \theta}\right) + \frac{\partial w}{\partial z} &= 0, && \text{(16d)}\\
\frac{DT}{dt} - K_H \nabla_H^2 T - K_V \frac{\partial^2 T}{\partial z^2} - \frac{(T_S - T^*)}{\tau_T}\mathcal{H}(\frac{z}{H_m} + 1) &= 0, && \text{(16e)}
\end{aligned}
$$

where $\mathcal{H}$ is a continuous approximation of the Heaviside function, g is the gravitational acceleration and $\tau_T = \rho_0 C_p H_m / \lambda_T$ is the surface adjustment time scale of heat (C_p is the constant heat capacity). In these equations, A_H and A_V are the horizontal and vertical momentum (eddy) viscosity and K_H and K_V the horizontal and vertical (eddy) diffusivity of heat, respectively. In addition, the operators in the equations above are defined as

$$
\begin{aligned}
\frac{D}{dt} &= \frac{\partial}{\partial t} + \frac{u}{r_0 \cos\theta}\frac{\partial}{\partial \phi} + \frac{v}{r_0}\frac{\partial}{\partial \theta} + w\frac{\partial}{\partial z},\\
\nabla_H^2 &= \frac{1}{r_0^2 \cos\theta}\left[\frac{\partial}{\partial \phi}\left(\frac{1}{\cos\theta}\frac{\partial}{\partial \phi}\right) + \frac{\partial}{\partial \theta}\left(\cos\theta \frac{\partial}{\partial \theta}\right)\right],\\
L_u(u,v) &= \nabla_H^2 u + \frac{u \cos 2\theta}{r_0^2 \cos^2\theta} - \frac{2\sin\theta}{r_0^2\cos^2\theta}\frac{\partial v}{\partial \phi},\\
L_v(u,v) &= \nabla_H^2 v + \frac{v \cos 2\theta}{r_0^2 \cos^2\theta} + \frac{2\sin\theta}{r_0^2\cos^2\theta}\frac{\partial u}{\partial \phi}.
\end{aligned}
\tag{17}
$$

Slip conditions and zero heat flux are assumed at the bottom boundary, while at all lateral boundaries no-slip and zero heat flux conditions are applied. As the forcing is represented as a body force over the first layer,

2Ω	=	$1.4 \cdot 10^{-4}$	$[s^{-1}]$	r_0	=	$6.4 \cdot 10^6$	$[m]$
H	=	$4.0 \cdot 10^3$	$[m]$	τ_T	=	$3.0 \cdot 10^1$	$[days]$
α_T	=	$1.0 \cdot 10^{-4}$	$[K^{-1}]$	g	=	9.8	$[ms^{-2}]$
A_H	=	$1.6 \cdot 10^5$	$[m^2 s^{-1}]$	T_0	=	15.0	$[^\circ C]$
ρ_0	=	$1.0 \cdot 10^3$	$[kgm^{-3}]$	A_V	=	$1.0 \cdot 10^{-3}$	$[m^2 s^{-1}]$
K_H	=	$1.0 \cdot 10^3$	$[m^2 s^{-1}]$	K_V	=	$1.0 \cdot 10^{-4}$	$[m^2 s^{-1}]$
C_p	=	$4.2 \cdot 10^3$	$[J(kgK)^{-1}]$	ΔT	=	20.0	$[K]$

Table 1. Standard values of parameters used in the minimal model.

slip and zero heat flux conditions apply at the ocean surface. Hence, the boundary conditions are

$$z = -H, 0 \quad : \quad \frac{\partial u}{\partial z} = \frac{\partial v}{\partial z} = w = \frac{\partial T}{\partial z} = 0, \tag{18a}$$

$$\phi = \phi_w, \phi_e \quad : \quad u = v = w = \frac{\partial T}{\partial \phi} = 0, \tag{18b}$$

$$\theta = \theta_s, \theta_n \quad : \quad u = v = w = \frac{\partial T}{\partial \theta} = 0. \tag{18c}$$

The parameters for the standard case are the same as in typical large-scale low-resolution ocean components of (4° horizontally) CGCMs and their values are listed in Table 1.

3.3 The AMO mode

The governing equations (16) and boundary conditions (18) are discretized on an Arakawa B-grid using central spatial differences. In the results in this section, a horizontal resolution of 4° is used. An equidistant grid with 16 levels is applied in the vertical so that the first layer thickness $H_m = 250$ m. The discretized system of equations on a $16 \times 16 \times 16$ grid with 5 unknowns per point (u, v, w, p and T) leads to a dynamical system of dimension, i.e., the number of degrees of freedom, $d = 5 \times 16^3 = 20,480$.

The steady equations are solved using a pseudo-arclength continuation method (Keller, 1977). As primary control parameter μ, we choose the equator-to-pole temperature difference ΔT. For every value of ΔT a steady solution of the minimal model is calculated under the restoring flux Q_{rest} in (13). For each steady flow pattern, the maximum of the meridional overturning streamfunction (ψ_M), where ψ is defined by

$$\psi(\phi, z, t) = -r_0 \cos\theta \int_{\phi_w}^{\phi_e} \int_{-H}^{z} v(\phi, \theta, z', t)\, dz' d\phi, \tag{19}$$

is calculated and it is plotted versus ΔT in Fig. 16a. The meridional overturning streamfunction for $\Delta T = 20$ K is plotted in Fig. 16b. The maximum of ψ occurs at about 60°N and the amplitude is about 16 Sv.

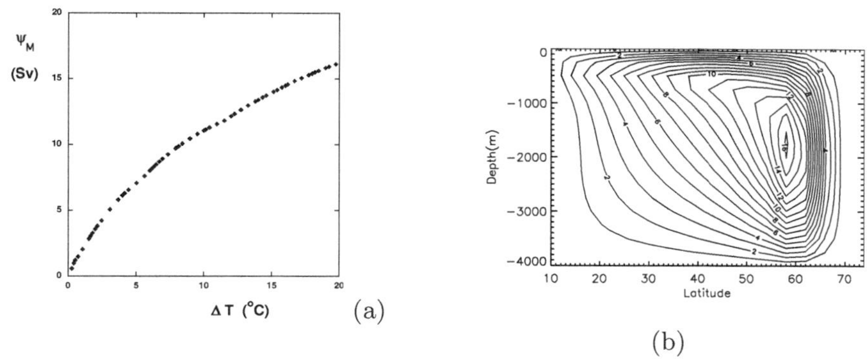

Figure 16. *(a) Plot of the maximum meridional overturning (ψ_M) of the steady solution (in Sv) versus the equator-to-pole temperature difference ΔT (°C) under restoring conditions. (b) Plot of the meridional overturning streamfunction (contour values in Sv) for $\Delta T = 20$ (°C).*

Next, the ocean-atmosphere heat flux Q_{pres} (where the subscript refers to 'prescribed') is diagnosed of each of the steady solutions and the linear stability of these solutions is computed under the heat flux Q_{pres}. The discretized linear stability problem of these steady states is formulated as a generalized eigenvalue problem

$$J\mathbf{x} = \sigma B\mathbf{x}. \tag{20}$$

Here $\sigma = \sigma_r + i\sigma_i$ where σ_r is the growth rate and σ_i the angular frequency, J is the Jacobian matrix and B is a singular matrix. We solve for the 'most dangerous' normal modes, i.e., those with σ closest to the imaginary axis. The growth rate and period $2\pi/\sigma_i$ of the mode with the largest growth rate are plotted versus ΔT in Fig. 17.

For $\Delta T = 20$°C, the AMO mode has a positive growth factor and hence the background state, of which the meridional overturning streamfunction was shown in Fig. 16b, is unstable to the AMO mode. The period of the AMO mode is about 67 years at $\Delta T = 20$°C, and it decreases with increasing ΔT (Fig. 17). From Fig. 17, we also see that the growth factor of the AMO mode decreases strongly with decreasing ΔT and becomes negative near $\Delta T_c \approx 4$°C, where a Hopf bifurcation occurs. For $\Delta T < \Delta T_c$, the steady states are therefore linearly stable under the prescribed flux Q_{pres}.

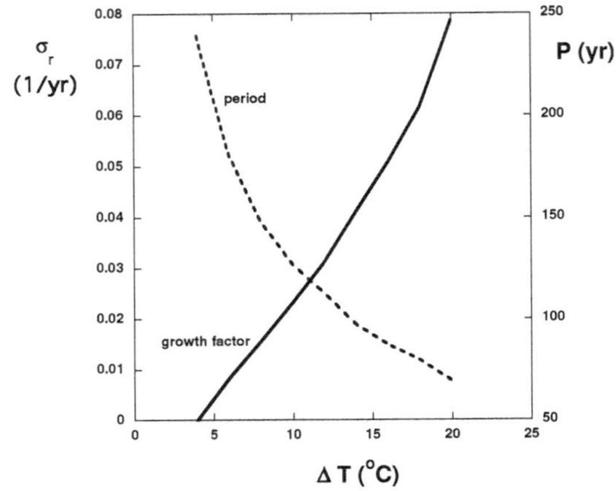

Figure 17. *Growth factor σ_r (in yr^{-1}, drawn) and period $P = 2\pi/\sigma_i$ (in yr, dashed) versus ΔT (in $^\circ C$) of the AMO mode in the minimal model under prescribed flux conditions.*

It was shown in Dijkstra (2006) that for small ΔT, the angular frequency of the AMO mode becomes zero and the complex conjugate pair of eigenvalues splits up into two real eigenvalues. The paths of the two different modes can be followed to the $\Delta T = 0°$C limit, where the eigensolutions connect to those of the diffusion operator of the temperature equation, called SST modes in Dijkstra (2006). These SST modes can be ordered according to their zonal (n), meridional (m) and vertical wavenumber (l) and it was found that the AMO mode connects to the $(0, 0, 1)$ SST mode and the $(1, 0, 0)$ SST mode near $\Delta T = 0°$C.

For each eigenvalue σ associated with the AMO mode, there is a corresponding eigenvector $\mathbf{x} = \mathbf{x}_r + i\mathbf{x}_i$ according to (20). In Fig. 18, the sea surface temperature field of the real part of the eigenvector ($\mathbf{x}_r$) of the AMO mode is plotted for $\Delta T = 4°$C (near the Hopf bifurcation) and for $\Delta T = 20°$C. With increasing ΔT, the pattern becomes more localized in the northwestern part of the basin.

3.4 Physical mechanism: the thermal Rossby mode

The physical mechanism of propagation of the AMO mode was presented in Colin de Verdière and Huck (1999) and Te Raa and Dijkstra (2002). This mechanism holds at every ΔT for which an oscillatory AMO mode is present (cf. Fig. 18). A sketch of this mechanism is provided in Fig. 19. A warm

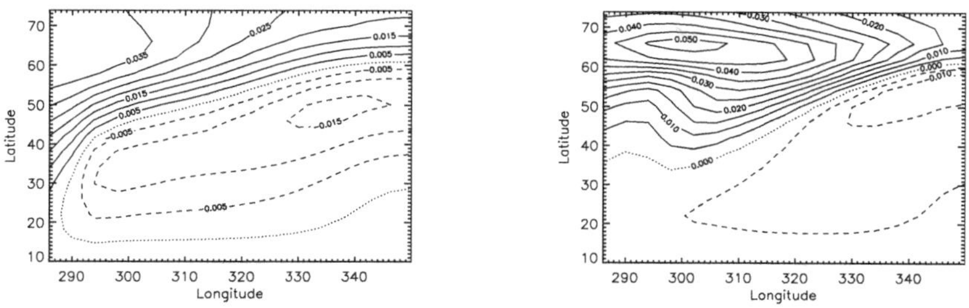

(a) (b)

Figure 18. *SST pattern of the real part the AMO mode for (a) $\Delta T = 4° C$ (near Hopf bifurcation) and (b) $\Delta T = 20° C$. Note that amplitudes are arbitrary as these are eigensolutions.*

anomaly in the north-central part of the basin causes a positive meridional perturbation temperature gradient, which induces – via the thermal wind balance – a negative zonal surface flow (Fig. 19a). The anomalous anticyclonic circulation around the warm anomaly causes southward (northward) advection of cold (warm) water to the east (west) of the anomaly, resulting in westward phase propagation of the warm anomaly. Due to this westward propagation, the zonal perturbation temperature gradient becomes negative, inducing a negative surface meridional flow (Fig. 19b). The resulting upwelling (downwelling) perturbations along the northern (southern) boundary cause a negative meridional perturbation temperature gradient, inducing a positive zonal surface flow, and the second half of the oscillation starts. The crucial elements in this oscillation mechanism are the phase difference between the zonal and meridional surface flow perturbations, and the westward propagation of the temperature anomalies (Te Raa and Dijkstra, 2002). The presence of freshwater forcing and wind stress forcing does not essentially change this mechanism; density anomalies will take over the role of temperature anomalies in the description above (Te Raa and Dijkstra, 2003; Te Raa et al., 2004).

Oscillatory behavior in the minimal model will occur for $\Delta T > \Delta T_c$ under prescribed flux conditions as the steady state is then unstable. A slight extension of minimal model is the inclusion of a more realistic continental shape. In Frankcombe et al. (2010), the same minimal model setup is used but with the horizontal resolution increased to $2° \times 2°$ and using 24 non-equidistant levels in the vertical with $H_m = 50$ m.

The variability is still concentrated in the northwestern part of the basin but the pattern has been deformed by the continents. To determine the

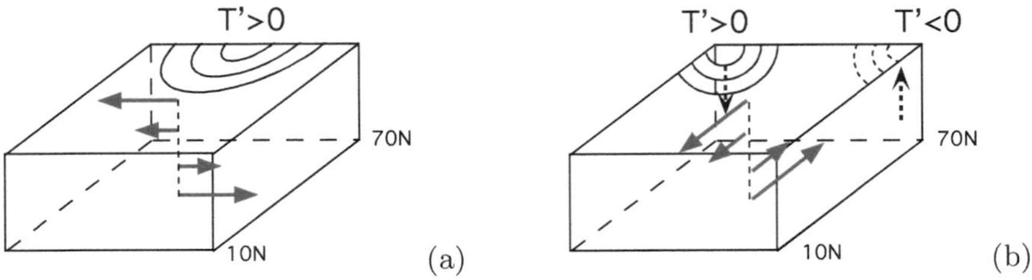

Figure 19. *Schematic diagram of the oscillation mechanism associated with the multidecadal mode caused by the westward propagation of the temperature anomalies T'. The phase difference between (a) and (b) is about a quarter period. See text and Te Raa and Dijkstra (2002) for a further explanation.*

AMO pattern as in Kushnir (1994), the difference in sea surface temperatures between the warm and cool phases of the AMO in the model is determined. In the model the SSTs from the year with the lowest AMO index were subtracted from those from the year with the highest AMO index. The resulting pattern is shown in Fig. 20. The overall effect of including

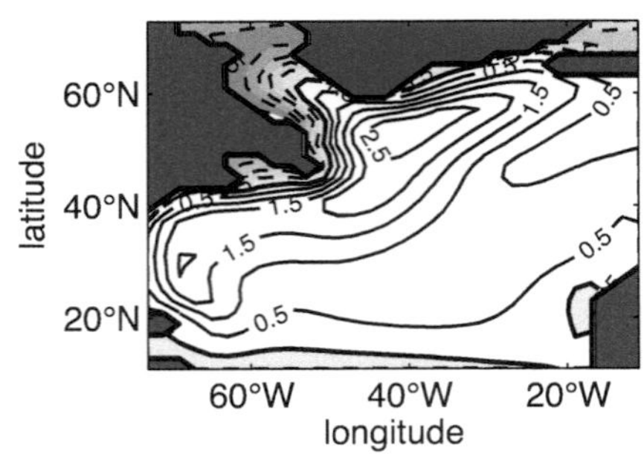

Figure 20. *The SST pattern of the AMO (determined as the difference between the patterns of SST from years with maximum and minimum AMO index) in the minimal model with continents; contour interval is 0.5° C.*

continents in the simple model is to deform the spatial pattern of the AMO mode into a form that more resembles observations (see Fig. 15).

4 Summary and Discussion

Using elementary dynamical systems theory and minimal models of ENSO and AMO, we have described basic mechanisms of variability behind both phenomena. We have shown that on an abstract level, on the dynamical systems level, both phenomena show the same characteristics. A steady background state becomes unstable through a Hopf bifurcation and the dominant mode of variability is related to a normal mode of the system. The spatial pattern of variability and the time scale are set by this normal mode. Obviously the physical processes of destabilization are very different for ENSO (coupled feedbacks) than for the AMO (thermal Rossby mode).

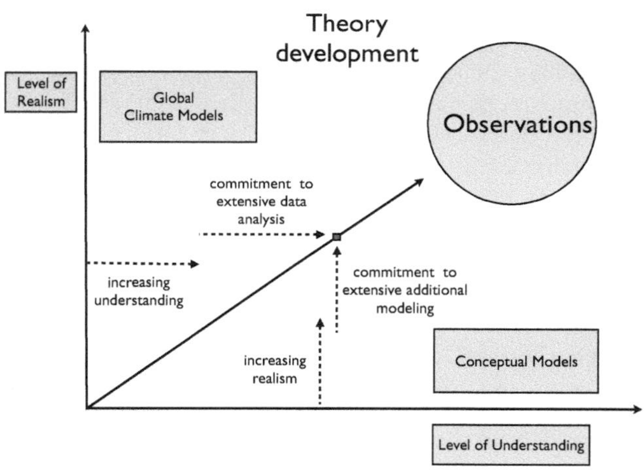

Figure 21. *Model usage for theory development on phenomena in the climate system.*

Obviously, observations are crucial for theory development as otherwise there is no way to falsify the proposed theories. However, theories based only on observations do not have to be consistent with the underlying conservation laws which must be satisfied in the system. Models are therefore always needed to interpret the observations and the whole model hierarchy (from conceptual models to GCMs) contributes to this interpretation. With the conceptual models, where the causal chains are precisely known, specific relations between quantities can be discovered, while with the GCMs the degree of realism (the quality of model-observation agreement) can be improved.

When one starts with a model high up in the hierarchy it may be difficult to connect the description of causal chains to well understood 'building blocks' as present in conceptual models. In this case, there must be a commitment to in-depth analysis of the model results. When starting with a model low in the hierarchy the connection to observations will be difficult. Hence, there must be a commitment to demonstrate that the processes identified in the conceptual model are also dominant in models higher up in the hierarchy (Fig. 21). Theories then develop in stages and towards the interpretation of the observations using successively better models of which the behavior is better understood.

One way to understand the processes driving a climate phenomenon using this approach is to start low in the hierarchy of deterministic models, say with a conceptual model as we did here with the minimal models. One identifies a basic mechanism consisting of the interaction of certain processes giving rise to the phenomenon. The next step is then to identify characteristics of this mechanism, so-called mechanistic indicators, which can be used to monitor whether this mechanism is present in models up in the model hierarchy. It is then crucial that the studies up to a model in the top of the hierarchy are really done and the properties of the mechanistic indicators investigated. When indeed the mechanism proposed from the conceptual model can be traced by its characteristics in a GCM and the latter provides an adequate simulation of observed phenomena, then we proceed into the right direction towards a theory of the phenomenon (Fig. 21).

For ENSO and the AMO, such a theory development is an active area of research for a long time already and many more details may be found in Dijkstra (2013).

Bibliography

D. Battisti and A. Hirst. Interannual variability in a tropical atmosphere-ocean model: Influence of the basic state, ocean geometry and nonlinearity. *J. Atmos. Sci.*, 46:1687–1712, 1989.

F Bretherton. *Earth system science: a closer view.* NASA, Washington DC, September 1988.

M. A. Cane and E. S. Sarachik. Forced baroclinic ocean motions: II. The linear equatorial bounded case. *Journal Of Marine Research*, 35:395–432, 1977.

F. Chen and M Ghil. Interdecadal variability in a hybrid coupled ocean-atmosphere model. *Journal Of Physical Oceanography*, 26:1561–1578, 1996.

Petr Chylek, Chris K Folland, Henk A Dijkstra, Glen Lesins, and Manvendra K Dubey. Ice-core data evidence for a prominent near 20 year time-scale of the Atlantic Multidecadal Oscillation. *Geophysical Research Letters*, 38:L13704, July 2011.

M. Claussen and coauthors. Earth system models of intermediate complexity: closing the gap in the spectrum of climate system models. *Climate Dynamics*, 18:579–586, March 2002.

A. Colin de Verdière and T. Huck. Baroclinic instability: an oceanic wavemaker for interdecadal variability. *Journal Of Physical Oceanography*, 29:893–910, 1999.

T. L. Delworth and R. G. Greatbatch. Multidecadal thermohaline circulation variability driven by atmospheric surface flux forcing. *J Climate*, 13:1481–1495, 2000.

T. L. Delworth and M. E. Mann. Observed and simulated multidecadal variability in the Northern Hemisphere. *Clim. Dyn.*, 16:661–676, 2000.

T. L. Delworth, S. Manabe, and R. J. Stouffer. Interdecadal variations of the thermohaline circulation in a coupled ocean-atmosphere model. *J Climate*, 6:1993–2011, 1993.

H A Dijkstra. Interaction of SST Modes in the North Atlantic Ocean. *Journal Of Physical Oceanography*, 36:286–299, 2006.

H A Dijkstra. *Nonlinear Climate Dynamics*. Cambridge University Press, Cambridge, 320 pp., 2013.

H A Dijkstra and G Burgers. Fluid Dynamics of El Nino Variability. *Annual Review of Fluid Mechanics*, 34:531–558, 2002.

H A Dijkstra and J.D. Neelin. Coupled ocean-atmosphere models and the tropical climatology. II: Why the cold tongue is in the east. *J Climate*, 8:1343–1359, 1995.

D. B. Enfield, A. M. Mestas-Nunes, and P. Trimble. The Atlantic multidecadal oscillation and its relation to rainfall and river flows in the continental US. *Geophys. Res. Letters*, 28:2077–2080, 2001.

A V Fedorov and S G Philander. Is El Niño Changing? *Science*, 288: 1997–2002, 2000.

Song Feng and Qi Hu. How the North Atlantic Multidecadal Oscillation may have influenced the Indian summer monsoon during the past two millennia. *Geophysical Research Letters*, 35:L01707, 2008.

L. M. Frankcombe, H A Dijkstra, and A S von der Heydt. The Atlantic Multidecadal Oscillation: a stochastic dynamical systems view. . In T. Palmer and P. Williams, editors, *Stochastic Physics and Climate Modelling*. Cambridge University Press, August 2010.

A. E. Gill. Some simple solutions for heat induced tropical circulation. *Quart. J. Roy. Meteor. Soc.*, 106:447–462, 1980.

R.J. Greatbatch and S. Zhang. An interdecadal oscillation in an idealized ocean basin forced by constant heat flux. *J Climate*, 8:82–91, 1995.

A. C. Hirst. Unstable and damped equatorial modes in simple coupled ocean-atmosphere models. *J. Atmos. Sci.*, 43:606–630, 1986.

T. Huck and G. Vallis. Linear stability analysis of the three-dimensional thermally-driven ocean circulation: Application to interdecadal oscillations. *Tellus*, 53A:526–545, 2001.

T. Huck, A. Colin de Verdi ere, and A. J. Weaver. Interdecadal Variability of the thermohaline circulation in box-ocean models forced by fixed surface fluxes. *Journal Of Physical Oceanography*, 29:865892, 1999.

T. Huck, G. Vallis, and Colin de Verdi ere. On the robustness of interdecadal oscillations of the thermohaline circulation. *J Climate*, 14:940–963, 2001.

N. Jiang, J.D. Neelin, and M Ghil. Quasi-quadrennial and quasi-biennial varibility in the equatorial Pacific. *Clim. Dyn.*, 12:101–112, 1995.

F.-F Jin. An equatorial recharge paradigm for ENSO. II: A stripped-down coupled model. *J. Atmos. Sci.*, 54:830–8847, 1997.

F.-F Jin and S I An. Thermocline and zonal advective feedbacks within the equatorial ocean recharge oscillator model for ENSO. *Geophys. Res. Letters*, 26:2989–2992, 1999.

F.-F Jin and J.D. Neelin. Modes of interannual tropical ocean-atmosphere interaction - a unified view. I: Numerical results. *J. Atmos. Sci.*, 50: 3477–3503, 1993.

Alexey Kaplan, Yochanan Kushnir, Mark A. Cane, and M. Benno Blumenthal. Reduced space optimal analysis for historical data sets: 136 years of Atlantic sea surface temperatures. *Journal of Geophysical Research*, 102(C13):27835–27860, 1997.

H. B. Keller. Numerical solution of bifurcation and nonlinear eigenvalue problems. In P. H. Rabinowitz, editor, *Applications of Bifurcation Theory*. Academic Press, New York, U.S.A., 1977.

R. A. Kerr. A North Atlantic climate pacemaker for the centuries. *Science*, 288:1984–1986, 2000.

Jeff R Knight, Chris K Folland, and Adam A Scaife. Climate impacts of the Atlantic Multidecadal Oscillation. *Geophysical Research Letters*, 33: L17706, 2006.

S. Kravtsov and M Ghil. Interdecadal Variability in a Hybrid Coupled OceanAtmosphereSea Ice Model. *Journal Of Physical Oceanography*, 34: 17561775, 2004.

Y. Kushnir. Interdecadal variations in North Atlantic sea surface temperature and associated atmospheric conditions. *Journal Of Physical Oceanography*, 7:141–157, 1994.

M. J. McPhaden and coauthors. The Tropical Ocean-Global Atmosphere observing system: a decade of progress. *Journal of Geophysical Research*, 103:14,169–14,240, 1998.

D. W. Moore. *Planetary-Gravity Waves in an Equatorial Ocean.* Harvard University, Cambridge, MA, U.S.A., 1968.

M. M. Munnich, M. Cane, and S. E. Zebiak. A study of self-excited oscillations of the tropical ocean-atmosphere system. II: Nonlinear cases. *J. Atmos. Sci.*, 48:1238–1248, 1991.

J.D. Neelin. The slow sea surface temperature mode and the fast-wave limit: Analytic theory for tropical interannual oscillations and experiments in a hybrid coupled model. *J. Atmos. Sci.*, 48:584–606, 1991.

J.D. Neelin and H A Dijkstra. Coupled ocean-atmosphere models and the tropical climatology. I: The dangers of flux-correction. *J Climate*, 8: 1325–1342, 1995.

J.D. Neelin, M. Latif, and F.-F Jin. Dynamics of coupled ocean-atmosphere models: The tropical problem. *Ann. Rev. Fluid Mech.*, 26:617–659, 1994.

J.D. Neelin, D. S. Battisti, A. C. Hirst, F.-F Jin, Y. Wakata, T Yamagata, and S. E. Zebiak. ENSO Theory. *Journal of Geophysical Research*, 103: 14,261–14,290, 1998.

J.D. Neelin, F.-F Jin, and H.-H Syu. Variations of ENSO phase locking. *J Climate*, 13:2570–2590, 2000.

S. G. H. Philander. *El Niño and the Southern Oscillation.* Academic Press, New York, 1990.

S. G. H. Philander, T Yamagata, and R. C. Pacanowski. Unstable air-sea interactions in the tropics. *J. Atmos. Sci.*, 41:604–613, 1984.

M. E. Schlesinger and N. Ramankutty. An oscillation in the global climate system of period 65-70 years. *Nature*, 367:723–726, 1994.

H. Stommel. Thermohaline convection with two stable regimes of flow. *Tellus*, 2:244–230, 1961.

M. Suarez and P S Schopf. A delayed action oscillator for ENSO. *J. Atmos. Sci.*, 45:3283–3287, 1988.

Rowan T Sutton and Daniel LR Hodson. Atlantic Ocean Forcing of North American and European Summer Climate. *Science*, 309:115–118, 2005.

L. A. Te Raa and H A Dijkstra. Instability of the thermohaline ocean circulation on interdecadal time scales. *Journal Of Physical Oceanography*, 32:138–160, 2002.

L. A. Te Raa and H A Dijkstra. Sensitivity of North Atlantic multidecadal variability to freshwater flux forcing. *J Climate*, 32:138–160, 2003.

L. A. Te Raa, J Gerrits, and H A Dijkstra. Identification of the mechanism of interdecadal variability in the North Atlantic Ocean. *Journal Of Physical Oceanography*, 34:2792–2807, 2004.

KE Trenberth. El Niño and climate change. *Geophysical Research Letters*, 24:3057–3060, 1997.

P. C. F. Van der Vaart, H A Dijkstra, and F.-F Jin. The Pacific Cold Tongue and the ENSO mode: Unified theory within the Zebiak-Cane model. *J. Atmos. Sci.*, 57:967–988, 2000.

C. Wang and J. Picaut. Understanding ENSO physics—A review. *Earth's Climate: The Ocean-Atmosphere Interaction. AGU Geophysical Monograph Series*, 147:21–48, 2004.

CZ Wang. A unified oscillator model for the El Nino-Southern Oscillation. *Journal Of Climate*, 14(1):98–115, 2001.

S. E. Zebiak and M. A. Cane. A model El Niño-Southern Oscillation. *Monthly Weather Review*, 115:2262–2278, 1987.

Rong Zhang and Thomas L Delworth. Impact of Atlantic multidecadal oscillations on India/Sahel rainfall and Atlantic hurricanes. *Geophysical Research Letters*, 33:L17712, 2006.

Y. Zhang, J. M. Wallace, and D. S. Battisti. ENSO-like interdecadal variability: 1900-1993. *J Climate*, 10:1004–1020, 1997.

A theoretical introduction to atmospheric and oceanic convection

Claudia Pasquero

Department of Earth and Environmental Sciences, University of Milan - Bicocca, Italy

Abstract The thermodynamical and dynamical bases for understanding the static stability and the conditional instability of atmospheric and oceanic fluid columns are here presented. The theoretical treatment is done in parallel for air and seawater, to highlight similarities and differences between the two. Vertical kinetic energy, as a measure of the intensity of convection, is discussed in terms of instability metrics.

1 Introduction

Atmospheric and oceanic convection is a very broad subject. Purpose of this chapter is not to condense in few pages a comprehensive discussion on known processes and open questions related to convection but rather to provide an introduction to the theory of the subject, following the outline of the lectures given by the author at the CISM summer school. We shall first introduce the relevant thermodynamic concepts, then discuss the momentum equation with emphasis on the energetics, and finally briefly discuss some applications. Atmospheric and oceanic cases are presented side by side, to highlight similarities and differences between the dynamics of the two geophysical fluids.

2 Thermodynamics

We here start with a detailed description of the thermodynamical properties of interest of the atmospheric and oceanic fluids. For more information on the material presented in this section, the reader is referred, among the many options available, to the inspiring book by Bohren and Albrecht (2009), to the lecture notes by Stevens, partially covered in Stevens (2005), to the book by Vallis (2006), and to the classical Emanuel (1994); Curry and Webster (1999); Gill (1982).

A. Provenzale et al. (Eds.), *The Fluid Dynamics of Climate*,
CISM International Centre for Mechanical Sciences
DOI 10.1007/ 978-3-7091-1893-1_2

2.1 Equation of state for dry and moist air

The air molecules can be considered as non interacting particles except during collisions, which are approximately elastic. Clearly, this assumption is not valid for studying aggregation of droplets and phase transitions, but when the different phases of water are in thermodynamical equilibrium it is a good approximation and the atmosphere can be idealized as a perfect gas. Given an air parcel, the specific mass of total water is the mass of water in it (in any phase) normalized to the total mass of the parcel[1] $q_t = \frac{m_w}{m_a} = \frac{m_v+m_l+m_i}{m_a} = q_v + q_l + q_i$, where for the last equality the definition of the specific mass of water in each phase has been used. Being α_v, α_l, α_i, α_d the specific volumes of water vapor, liquid water, ice, and dry air respectively, the volume occupied by the mass m_a of air is $V = m_v\alpha_v + m_l\alpha_l + m_i\alpha_i \simeq m_v\alpha_v = m_d\alpha_d$, as liquid water and ice occupy a negligible volume with respect to water vapor, and for the last equality we have used the consideration that the same volume is shared between water vapor and dry air. It follows that the specific volume of the mixture can be written as

$$\alpha = \alpha_v q_v = \alpha_d q_d, \tag{1}$$

with q_d being the specific mass of dry air.

We now write the ideal gas equation of state, $p = \rho RT$ (where $R = R^*/M$ is the universal gas constant divided by the molar mass of the gas considered, p is pressure, $\rho = \alpha^{-1}$ is density, and T is temperature) for water vapor and dry air separately: using eq.1, partial pressure of water vapor is $e = \alpha_v^{-1} R_v T = \rho q_v R_v T$ and partial pressure of dry air is $p_d = \alpha_d^{-1} R_d T = \rho q_d R_d T$. Dalton's law allow us to write the pressure of moist air as the sum of the partial pressures of water vapor and of dry air, $p = e + p_d = \rho(R_v q_v + R_d q_d)T$, or, equivalently,

$$p = \rho R_d(1 + \epsilon q_v - q_l - q_i)T, \tag{2}$$

with $\epsilon = (M_d - M_w)/M_w \simeq 0.608$, being M_d and M_w the molar mass of dry air and of water, respectively.

The sum of all the specific masses at play is equal to one by construction, $q_v + q_l + q_i + q_d = 1$ and eq. 2 can be written in the general form $p = \rho RT$, where the gas constant depends on the water content through

$$R = R_d(1 + \epsilon q_v - q_l - q_i) \tag{3}$$

or, alternatively, as $p = \rho R_d T_\rho$ defining the *density temperature* T_ρ:

$$T_\rho = T(1 + \epsilon q_v - q_l - q_i).$$

[1]The adjective *specific* in this chapter indicates that the referenced quantity is normalized to the mass of the fluid considered, either air or seawater

In absence of condensate, the density temperature corresponds to the *virtual temperature*, which is defined as the temperature that dry air would have to have at pressure p in order to have the same density of moist air at temperature T and pressure p.

For future reference, we also note that following a similar procedure the specific heat capacities of mixtures of ideal gases can be derived, as

$$c_p = c_{pd}q_d + c_{pv}q_v \tag{4}$$

and similarly for c_v.

2.2 Equation of state for seawater

Interactions between molecules in liquid water cannot be neglected and seawater is never well approximated as an ideal gas. Rather than being derived from first principles, the UNESCO seawater equation of state (IES 80, Fofonoff (1995)) is a complex analytical expression that fits very precise measurements. We here write it as

$$\rho = \rho(S, T, p) = \rho_0 \left[1 - \beta_T(T - T_0) - \beta_S(S - S_0) - \beta_p(p - p_0)\right].$$

The values of the β parameters (called thermal expansion coefficient, saline contraction coefficient, and compressibility coefficient) vary significantly in the ocean, with temperature, salinity, and pressure. In particular, the thermal expression coefficient has the largest relative variations, as it significantly increases both with temperature and pressure. For our discussion, it is important to consider the dependence of β_T on pressure, and we shall use $\beta_T(p) = \beta_T^o[1 + \gamma^o(p - p_0)]$, where the superscript indicates a constant and γ^o explicits the *thermobaricity* of the system. The values of β_S and β_p will be considered constant, so that the seawater equation of state that will be used in the following is:

$$\rho = \rho_0 \left\{1 - \beta_T^o[1 + \gamma^o(p - p_0)](T - T_0) - \beta_S^o(S - S_0) - \beta_p^o(p - p_0)\right\}. \tag{5}$$

Notice that even if this approximated equation is valid only near its reference state, it captures the nonlinearities that are important for our discussion.

2.3 Potential temperature and adiabatic lapse rates for a sub-saturated atmosphere

The thermodynamic equation is here written following the notation of Bohren and Albrecht (2009), who abandon the common use of differentials and consider instead the temporal rate of change of the thermodynamic quantities. For instance, rather than writing the first law of thermodynamics as $du = dq - dw + dc$, we here write a tendency equation for the specific internal energy u of a fluid parcel as the sum of work rate done by or on it, w, heat input at a rate $\dot{q}$, and/or a change rate in its chemical composition c:

$$\frac{Du}{Dt} = \dot{q} - w + \frac{Dc}{Dt}. \tag{6}$$

From the definition of work as the path integral of the product of the force F exerted on a mass in the direction of the displacement and this displacement dl, and considering that pressure is the force normal to a surface divided by the surface area, A, we get the specific work rate as the pressure multiplied by rate of change of the specific volume, i.e. $w = p\frac{D\alpha}{Dt}$. In the following we shall neglect changes in chemical composition of the fluid, $\frac{Dc}{Dt} = 0$. For an ideal gas and for an isocore transformation, eq.6 implies $\frac{Du}{Dt} = \dot{q} = \frac{Dq}{Dt} = \left(\frac{dq}{dT}\right)_v \frac{DT}{Dt} = c_v\frac{DT}{Dt}$. Internal energy is a state function so the equation $\frac{Du}{Dt} = c_v\frac{DT}{Dt}$, although derived for a specific transformation, can be used for any transformation. The first law of thermodynamics for an ideal gas of constant chemical composition is thus

$$c_v\frac{DT}{Dt} + p\frac{D\alpha}{Dt} = \dot{q}$$

from which, writing $p\frac{D\alpha}{Dt} = \frac{D(p\alpha)}{Dt} - \alpha\frac{Dp}{Dt}$ and making use again of the ideal gas equation of state $\alpha p = RT$ and recalling that $c_p = R + c_v$, we obtain

$$c_p\frac{DT}{Dt} - \alpha\frac{Dp}{Dt} = \dot{q}. \tag{7}$$

For a reversible process the specific heating rate can be written in terms of the specific entropy s as

$$\dot{q} = T\frac{Ds}{Dt}. \tag{8}$$

Eq.7 becomes

$$\frac{Ds}{Dt} = c_p\frac{1}{T}\frac{DT}{Dt} - \frac{\alpha}{T}\frac{Dp}{Dt} = \frac{c_p}{T}\frac{DT}{Dt} - \frac{R}{p}\frac{Dp}{Dt} = \frac{D}{Dt}(c_p \ln T - R\ln p) \tag{9}$$

Note that to write the last equality in the above equation we assumed that the specific heat coefficients do not change in time. To this end, recall that in section 1 we derived the specific heat coefficients and the gas constant for a mixture of different components, and we noted that they depend on the specific masses. We already restricted our theory to the case of constant chemical composition of the fluid, but we now need to add a further restricion as equation 9 is valid only if there are no phase transitions, such that

R and c_p are indeed constant. The cases in which phase transitions occur must be treated separately, and will be done in the next section.

Integration of eq.9 from a state characterized by T, p and s to a state T_0, p_0, s_0, leads to the expression of specific entropy variations as

$$s_0 - s = c_p \ln \frac{T_0}{T} - R_d \ln \frac{p_0}{p}. \tag{10}$$

If the transformation is isentropic, the temperature T_0 in the above equation assumes the name of *potential temperature*, θ. In other words, θ is the temperature that the gas at temperature T and pressure p would get when reaching pressure p_0 under an adiabatic transformation. Substituting θ for T_0 and $s = s_0$ in the above equation leads to:

$$\theta = T \left(\frac{p_0}{p} \right)^{R/c_p}. \tag{11}$$

The term *potential density* refers to the density that the fluid with potential temperature θ has at the reference pressure p_0: $\rho_\theta = \rho(\theta, p_0) = \frac{p_0}{R\theta}$.

The first law of thermodynamics following a fluid parcel, eq. 9, can be rewritten as

$$c_p \frac{D\theta}{Dt} = \frac{\theta}{T} \dot{q}.$$

The diabatic heating term $\dot{q}$ on the r.h.s. of the above equation includes radiative heating and heat diffusion.
It is immediate to verify, using eq.8, that $\frac{Ds}{Dt} = \frac{c_p}{\theta} \frac{D\theta}{Dt}$, which integrated leads to the relation between entropy and potential temperature: $s = c_p \ln \theta + \text{const}$. An isentropic transformation is thus caracterized by constant θ, where θ is calculated with the appropriate values of c_p and R. Note also that $\frac{D\theta}{Dt} = 0$ implies that potential density is conserved as well, as long as the value of R does not change (which means that in absence of phase transitions potential density is conserved for adiabatic transformations).

To get the adiabatic lapse rate, i.e. the rate at which the temperature of a parcel changes in the vertical undergoing an adiabatic displacement, it is sufficient to substitute $\dot{q} = 0$ in the equation above, and obtain $\frac{D\theta}{Dt} = \frac{\partial\theta}{\partial t} + \mathbf{u} \cdot \nabla\theta = 0$. Considering $u = v = 0$ and θ independent of time we find $\frac{d\theta}{dz} = 0$, from which, using eq. 11,

$$0 = \frac{d}{dz} \left[T \left(\frac{p_0}{p} \right)^{R/c_p} \right] = \left(\frac{p_0}{p} \right)^{R/c_p} \frac{dT}{dz} - T \left(\frac{p_0}{p} \right)^{R/c_p} \frac{R}{c_p} \frac{1}{p} \frac{dp}{dz}$$

and, using the hydrostatic equation $\frac{dp}{dz} = -\rho g$ and the ideal gas equation of state once more, we find the adiabatic lapse rate

$$\frac{dT}{dz} = \Gamma = -\frac{g}{c_p}. \tag{12}$$

We recall here that the above treatment is valid for an ideal gas with constant R and c_p. It is thus valid for the atmosphere, provided the correct values of R and c_p are used (which depend on the chemical composition and on the water content, see eq.3 and eq.4) and assuming there are no phase transitions, nor changes in its chemical composition. For this reason, the air parcel must be undersaturated during all the displacement. The lapse rate in eq.12 is referred to as the dry adiabatic lapse rate, despite it can account for moisture in the definition of R and c_p, because there is no liquid nor solid water in the subsaturated parcel. Strictly speaking, we should refer to the dry adiabatic lapse rate when we consider $\Gamma_d = -g/c_{pd}$, and to the moist adiabatic lapse rate when we consider $\Gamma_m = -g/(c_{pd}q_d + c_{pv}q_v)$. Given that $c_{pv} > c_{pd}$, we get a value of Γ_m smaller than Γ_d but considering that the values of q_v found in the atmosphere are at most O(0.01), the difference between the two adiabatic lapse rates is negligible. Please notice that the term moist adiabatic lapse rate is usually not referred to Γ_m, but rather to the saturated adiabatic lapse rate that we'll discuss in the following.

2.4 Potential temperature and adiabatic lapse rates for a saturated atmosphere

We now focus on the case in which phase transitions of water between the gaseous and the liquid phase are allowed, and consider the specific entropy of a mixture of gases (dry air and moisture) and liquid water. We consider the transition from a state characterized by a temperature T, pressure p and specific mass of dry air q_d, specific mass of total water q_t, subdivided into the specific mass of liquid water q_l and of water vapor $q_v = (q_t - q_l)$, to a state at temperature T_0, pressure p_0, and specific mass of water vapor $q_{v0} = 0$, meaning that all water is in the liquid phase. A different reference state could be defined, such as for instance the state in which the total water mass is in the gaseous phase up to the saturation level, and the rest is in the liquid phase. We shall use such a definition later on, and here stick to the common practice of using all water in the liquid phase at the reference state. For conservation of mass of dry air and of water, q_d and q_t do not change during the transition. Total specific entropy is given by the weighted sum of the specific entropy of the individual constituents (dry air, water vapor, liquid water):

$$s = q_d s_d + q_l s_l + q_v s_v = q_d s_d + q_t s_l - q_v(s_l - s_v),$$

while in the reference state the specific entropy is

$$s_0 = q_d s_{d0} + q_t s_{l0}.$$

If the transformation is isentropic, the reference temperature T_0 assumes the name of *equivalent potential temperature*, θ_e, the density of air with θ_e at pressure p_0 is the *equivalent potential density*, $\rho_{\theta_e} = \rho(\theta_e, p_0)$, and $s = s_0$, i.e.

$$q_d(s_d - s_{d0}) + q_t(s_l - s_{l0}) - q_v(s_l - s_v) = 0. \tag{13}$$

We now aim at writing the three addenda in the above sum in terms of the thermodynamic variables specified above. The partial pressure of dry air in the starting state can be expressed in terms of the defined variables using eq.2:

$$p_d = \frac{pq_d}{(1 + \epsilon q_v - q_l)}$$

while in the final state, considering that the reference state has no water vapor, the partial pressure of dry air is equal to the total pressure p_0. During the transition, dry air changes temperature from T to θ_e and pressure from the partial pressure p_d to p_0 such that, from eq.10, we get the change in specific entropy for dry air:

$$s_d - s_{d0} = c_{pd} \ln \frac{T}{\theta_e} - R_d \ln \frac{p_d}{p_0}.$$

For liquid water, assuming incompressibility, we have $\dot{q} = c_{pl} \frac{DT}{Dt}$, where c_{pl} is the specific heat of liquid water at constant pressure[2], and thus

$$\frac{Ds_l}{Dt} = \frac{\dot{q}}{T} = c_{pl} \frac{1}{T} \frac{DT}{Dt}. \tag{14}$$

Integration of this equation from temperature T to temperature θ_e leads to

$$s_l - s_{l0} = c_{pl} \ln \frac{T}{\theta_e}. \tag{15}$$

The last term in eq.13 is further written as $q_v(s_v - s_l) = q_v(s_v - s^* + s^* - s_l)$, where s^* is the specific entropy of water vapor at temperature T and

[2]for liquids, the specific heat at constant pressure and at constant volume are nearly identical, and usually referred to as heat capacity. We prefer to keep the reference to constant pressure in the notation for consistency with the gaseous phase.

saturation pressure e^*. We can thus use eq.10 to obtain the change in specific entropy for water vapor from temperature T and pressure $e = \frac{pq_v}{1+\epsilon q_v - q_l}$ to temperature T and pressure e^*:

$$(s_v - s^*) = -R_v \ln \frac{e}{e^*}.$$

The specific entropy change from water vapor at saturation conditions to liquid water at constant temperature is the specific enthalpy difference between the gaseous and the liquid phases (i.e. the specific latent heat of condensation) divided by the temperature T at which the transition occurs:

$$(s^* - s_l) = \frac{l_v}{T}.$$

We are finally ready to rewrite eq.13 as

$$(q_d c_{pd} + q_t c_{pl}) \ln \frac{T}{\theta_e} - q_d R_d \ln \frac{p_d}{p_0} - q_v R_v \ln \frac{e}{e^*} + q_v \frac{l_v}{T} = 0.$$

We then define $c_{p0} = q_d c_{pd} + q_t c_{pl}$ and solve for θ_e:

$$\theta_e = T \left[\frac{p_0}{p} \frac{(1 + \epsilon q_v - q_l)}{q_d} \right]^{\frac{q_d R_d}{c_{p0}}} \left(\frac{e}{e^*} \right)^{-\frac{q_v R_v}{c_{p0}}} \exp \left(\frac{q_v l_v}{c_{p0} T} \right). \tag{16}$$

Note that the dependence on q_l enters explicitly in the term in square brackets.

When air is moved vertically with no exchange of heat with the surrounding, the value of θ_e must be conserved, therefore the adiabatic lapse rate is obtained imposing that $\frac{d\theta_e}{dz} = 0$. The quantities on the r.h.s. of equation 16 that depend on height z are temperature T, pressure p, and specific masses of water. To express the adiabatic lapse rate of temperature as function of dp/dz, which is known under the hydrostatic approximation as function of gravity and density, we need to specify how the water vapor specific mass changes with height. An assumption is thus needed at this point. By assuming that the moist air parcel is always as saturation, $q_v = q_s$, and using the hydrostatic equation and the Clausius-Clapeyron equation, the *saturated adiabatic lapse rate* can be derived, $\frac{\partial T}{\partial z} = -\frac{g}{c_p^*}$. The effective isobaric heat capacity

$$c_p^* = \frac{c_{p0} \frac{R}{q_d R_d} + \frac{q_v}{R_v} \left(\frac{R L_v}{q_d R_d T} \right)^2}{1 + \frac{q_v L_v}{q_d R_d T}}$$

is a function of temperature T and depends also on the specific mass of water vapor q_v which enters both implicitly in the definition of R, and explicitly. The value of c_p^* is thus dependent on the height at which it is computed. Temperature T following an adiabatic displacement of air at saturation does not change linearly with height. Note however that this nonlinearity is not crucial for what will be discussed in the following sections, as it does not qualitatively change the stability of the air columns. The important fact is that this effective isobaric heat capacity is larger than its dry counterpart, yelding a smaller adiabatic lapse rate for saturated air than for subsaturated air.

A further simplification is often made, in which liquid water is assumed to immediately precipitate form the air parcel, such that $q_l = 0$ during the transformation. Clearly, such a situation is not reversible nor strictly adiabatic, as rain carries heat away: entropy is not conserved. However, the change in entropy is small and is often neglected. The *pseudo adiabatic lapse rate* is obtained by imposing $q_l = 0$ in the expression for c_p^*.

Natural atmospheric processes are neither saturated adiabatic nor pseudoadiabatic, but somewhere in between. The numerical difference between the two is however very small and the two terms are often used as synonimous.

Before concluding this section, we want to define, for later use, a different type of equivalent potential temperature, in which in the reference state all water is supposed to be in the gaseus phase if below saturation, and divided between the gaseous and the liquid phase if above saturation, with relative humidity equal to 1. Following a procedure similar to the one presented above, this type of potential temperature is found to be

$$\theta_{e*,p} = T\left[\frac{p_0}{p}\frac{(1+\epsilon q_v - q_l)}{q_d}\right]^{\frac{q_d R_d}{c_{p0}}}\left(\frac{e}{e^*}\right)^{-\frac{q_v R_v}{c_{p0}}}\exp\left(\frac{(q_v - q_v^*)l_v}{c_{p0}T}\right).$$

Analogously, we can define the corresponding equivalent potential density $\rho_{\theta,e^*,p}$.

2.5 Potential temperature and adiabatic lapse rate for the ocean

To simplify our treatment we shall not consider salinity variations in the following. Under such assumption, entropy depends on temperature and pressure only. Following the reasoning that allowed us to write eqs.14 and 15, we can immediately derive that potential temperature, defined as the temperature that a water parcel would get when adiabatically displaced to a reference pressure p_0, is conserved during adiabatic transformations, exactly as for an ideal gas of constant composition, as we get

$$\frac{Ds}{Dt} = \frac{c_{pl}}{\theta}\frac{D\theta}{Dt}.$$

Note that, unlike for an ideal gas of constant composition, the specific heat of seawater depends on salinity, temperature, and pressure. The main dependence is on salinity so that, given that we hare note considering salinity changes, we shall keep c_{pl} as constant in the following. The above equation, valid in the case of constant salinity as otherwise entropy should depend on chemical composition as well, is approximately valid for the ocean, as saline source terms and saline diffusion are small. It implies that potential temperature is conserved during adiabatic transformations. We have not however specified how potential temperature depends on temperature, salinity, and pressure. The procedure is explicited in Vallis (2006), chapter 1.8, and leads to

$$\theta = T\exp\left\{-\frac{\beta_T^o(p-p_0)}{\rho_0 c_{pl}}\left[1+\frac{1}{2}\gamma^o(p-p_0)\right]\right\}.$$

Potential density is the density that the water parcel would get when moved adiabatically and without changing salinity to the reference pressure p_0, i.e. $\rho_\theta = \rho_0\left[1-\beta_T^o(\theta-T_0)\right]$, whose approximate validity is guaranteed by the fact that potential temperature is quite similar to temperature so the equation is used in the vicinity of the reference state, as long as T is close to T_0.

Imposing now conservation of potential temperature along an adiabatic vertical displacement, $\frac{d\theta}{dz}=0$, and using the hydrostatic equation, allows us to write the adiabatic lapse rate as

$$\frac{dT}{dz} = -\beta_T^0 T\frac{g}{c_{pl}}[1+\gamma^0(p-p_0)]. \tag{17}$$

It is important to note that there is an explicit dependence on pressure in the above equation. This means that the value of the adiabatic lapse rate changes with depth. This peculiarity has te be kept in mind for the discussion in the next section.

3 Dynamics

We here discuss the conditions that lead to motion in the vertical direction. To this end, static stability is first introduced for both the atmosphere and the ocean, and then conditional instability is presented for the two fluids. The concept of conditional instability has been largely used in the atmospheric context but is much less developed in its oceanic applications and probably deserves more attention (Su et al. (2012)). We here present the atmospheric and oceanic conditional instability together to highlight their similarities and their differences.

3.1 Inviscid static stability

We now consider a fluid parcel of volume V located at height z with density corresponding to the environmental density $\rho(z) = \tilde{\rho}(z)$ (here the tilde variables denote the environment conditions and the undecorated variables refer to the parcel). The parcel is vertically displaced adiabatically to height $z + z'$ where it has density $\rho(z + z')$ in an environment with density $\tilde{\rho}(z + z')$. The buoyancy force acting on it is $-(\rho(z + z') - \tilde{\rho}(z + z'))Vg$. Now, choose the pressure at $(z + z')$ as the reference value to define the potential density, such that potential density and density at height $(z + z')$ are equivalent, i.e. $\rho(z+z') = \rho_\theta(z+z')$ and $\tilde{\rho}(z+z') = \rho_\theta(z) = \tilde{\rho}_\theta(z+z')$. This choice is actually a key point. The acceleration on the parcel is thus

$$\frac{d^2 z'}{dt^2} = -\frac{\rho_\theta(z + z') - \tilde{\rho}(z + z')}{\rho_\theta(z + z')} g. \tag{18}$$

We further note that, in absence of phase transitions, potential density is conserved upon an adiabatic displacement, $\rho_\theta(z + z') = \tilde{\rho}(z)$, and, considering a small displacement z' so that the environmental potential density can be linearized, we get

$$\frac{d^2 z'}{dt^2} = \frac{g}{\tilde{\rho}_\theta} \frac{d\tilde{\rho}_\theta}{dz} z' = -\frac{g}{\tilde{\theta}} \frac{d\tilde{\theta}}{dz} z'. \tag{19}$$

Such equation is the oscillator equation with the Brunt-Väisälä frequency

$$N^2 = -\frac{g}{\tilde{\rho}_\theta} \frac{d\tilde{\rho}_\theta}{dz}.$$

It implies that the condition for stability of parcels to vertical displacements is that $N^2 > 0$, i.e. $\frac{d\tilde{\rho}_\theta}{dz} < 0$, or the environmental potential density decreases with height. This is true both for dry and moist but subsaturated air, provided the correct value of c_p is used. In the case of dry air the condition can be rewritten as $\frac{d\tilde{\theta}}{dz} > 0$, or $\frac{d\tilde{T}}{dz} > -\Gamma_d$, meaning that if the environmental temperature decreases slower with height than does the dry adiabatic lapse rate the atmosphere is statically stable. For moist subsaturated air the condition on the temperature lapse rate must account for possible changes with height of the environmental specific mass of water vapor. Rather than writing the exact threshold for the environmental lapse rate, for the sake of generality we prefer to refer to the condition on the environmental potential density.

In the case of saturated air, the equivalent potential density must be used instead of the potential density and the stability depends on the sign of

$$N^2 = -\frac{g}{\tilde{\rho}_{\theta_e}} \frac{d\tilde{\rho}_{\theta_e}}{dz}.$$

When expressing it in terms of the environmental temperature lapse rate, the condition depends on the saturated adiabatic lapse rate, on the vertical rate of change of water vapor specific mass, and on the latent heat released or absorbed during condensation and evaporation of water vapor, to keep the air parcel at saturation.

The condition for static stability in the ocean is the same as for the atmosphere, based on the sign of the environmental potential density profile, keeping in mind that it depends on the thermal profile and on the salinity content at each level. For given temperature and salinity, the adiabatic lapse rate depends on pressure (eq.17) and thus a water column with a given environmental lapse rate of density might be statically stable at some pressure and unstable at some other pressure.

The instability conditions defined on the sign of the environmental lapse rate (a.k.a. Schwarzschild criterion, as it was first described in Schwarzschild (1906)) is a necessary condition, which is however not sufficient for the development of convection. Dissipative factors, such as viscosity and thermal conduction, inhibit the convective instability and may prevent vertical motion all together even in presence of static instability in the fluid.

Both the atmosphere and the ocean, except for the first layers close to the surface, are typically characterized by positive N^2, i.e. they are statically stable. However, in both fluids deep convective motion can be set up, and this will be discussed in the next section.

3.2 Conditional instability

The atmosphere is often *conditionally* unstable, meaning that the instability is conditioned to a large enough displacement. A subsaturated parcel adiabatically displaced vertically by a small distance typically oscillates around its original position, but if it is displaced by a larger distance its temperature might decrease (following a dry adiabat) to the point the parcel becomes saturated (the level at which this happens is the lifting condensation level, LCL). From that level up, the temperature of the parcel will decrease nearly following a saturated adiabat, and, if still lifted up, can reach a level at which the parcel is warmer than the environment. At that level, called the level of free convection (LFC), the parcel will begin to accelerate upwards and no work needs to be done to move it further up. Note that at the level of free convection the atmosphere is statically stable: a parcel at that level is stable to small displacements. This is due to the fact that air at that level is much drier than air that is adiabatically lifted from lower levels, so the displaced parcel follows a dry adiabat.

We now want to look at the conditional instability from a slightly different point of view. The stability condition based on the potential density lapse rate has been constructed using the level z to choose the reference pressure for the definition of potential density. The reference pressure is thus no longer arbitrary. In atmospheric sciences, however, the reference pressure is usually considered the surface pressure, p_s, i.e. the pressure at $z = 0$. Let us see why that is the case, calling ρ_{θ,p_s} the potential density calculated using the surface pressure as reference pressure, and ρ_{θ,p_z} the potential density calculated using pressure $p_z = p(z)$ as reference pressure. If we want to know whether surface air is stable in a given environment, we typically ask whether the potential density $\rho_{\theta,p_s}(0)$ of surface air is larger or smaller than the potential density $\rho_{\theta,p_s}(z)$ of air at a given level z. To be precise, one should ask whether the density that surface air would have if lifted to the level z (which corresponds to $\rho_{\theta,p_z}(0)$) is larger or smaller than the in situ density at level z (which by definition corresponds to the potential density $\rho_{\theta,p_z}(z)$). The two methods however give the same answer, provided no phase transitions would occur, justifying the use of meteorologists of surface pressure as reference pressure. However, we here prefer to stick to the more general criterion for stability of a surface air parcel displaced to height z, and compare the environmental density profile with the density profile obtained by adiabatically lifting a surface air parcel, assuming that the specific mass of water vapor at each level is capped by the local saturation value. Below the level at which surface air reaches saturation, we simply compare $\tilde{\rho}(z)$ with $\rho_{\theta,p_z}(0)$ for the surface air: $\rho_{\theta,p_z}(T^s, q_v^s)$, where T^s and q_v^s are surface air temperature and surface specific mass of water vapor. Above the LCL, we compare $\tilde{\rho}(z)$ with $\rho_{\theta,e^*,p_z}(0)$, where equivalent potential density is calculated using as reference values pressure $p(z)$, and assuming that the parcel is saturated, with specific mass of water vapor q_v^*. The two profiles obtained are shown in fig.1, for an undersaturated atmosphere with temperature lapse rate equal to 7 K/km, with surface conditions $T^s = 25\,\deg C$ and $q_v^s = 0.01 kg/kg$. There are several nonlinearities that enter in the density profile that an air parcel lifting from surface pressure to a reference pressure p would get if raised adiabatically

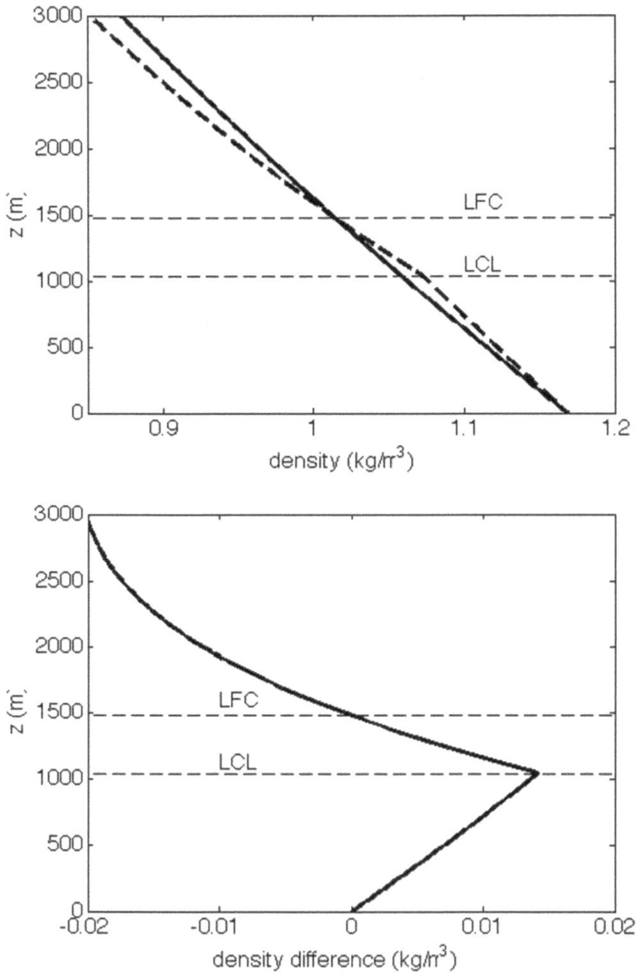

Figure 1. The top panel shows (solid line) the environmental density for an unsaturated atmosphere with a surface temperature of 25°C and a thermal lapse rate of 6 K/km, and (dashed line) the density that surface air with 60% relative humidity would have if adiabatically lifted to level z and all water vapor in eccess of saturation at that level condensated. The bottom panel shows the difference between the two curves of the top panel. Positive values indicate that surface air lifted at the given level would be denser than environmental air. Labels indicate the lifting condensation level (LCL) and the level of free convection (LFC).

(or pseudo-adiabatically), but the really important non linearity to determine wheter surface air is conditionally unstable comes from the fact that at the lifting condensation level adiabatic density lapse rate changes from dry adiabatic to saturated adiabatic, as visible in fig. 1.

The ocean, as the atmosphere, is typically statically stable with respect to infinitesimal perturbations, except for the top layers where shallow convection occurs and for a handful of location around the globe where deep convection takes place (polar regions and the Gulf of Lion in the Mediterranean). However, as for the atmosphere, it can be conditionally unstable. Given that the thermal expansion coefficient increases with pressure, the temperature difference between two water masses at large depths leads to a larger density difference than at the surface. A stably stratified ocean can be conditionally unstable if, for instance, bottom water is slightly colder and fresher than the water above. In fig.2 we choose to show the conditonal instability for bottom water, to allow a direct comparison with the more familiar conditional instability for surface air in the atmosphere. To this aim, we have used the simplified equation of state for seawater, eq.5 and the corresponding adiabatic lapse rate, eq.17. It is clear from this analysis that the thermobaric term, γ^0, is responsible for the curvature of the adiabatic density lapse rate which allows the crossing of a profile with a constant environmental density lapse rate.

Conditional instability in the ocean is found in the Weddell Sea (Akitomo (1999a); McPhee (2003), and see fig.3) and might play an important role in the deep water formation process and possibly in climate variability (Adkins et al. (2005)). In the Weddell Sea, mixed layer water is typically fresher and colder than the relatively homogeneous deep water: at low pressure the salinity difference dominates and the saltier water below the mixed layer is denser than the fresher mixed layer water, leading to a statically stable water column. However, if mixed layer water was brought to a large enough pressure, the temperature difference between the two water masses would dominate, and the mixed layer colder water would be denser than the warmer deep water. It is believed that during some winters anomalous cooling of the upper ocean initiates convection that is strengthened by thermobaricity (Akitomo (1999b)).

In summary, conditional instability in the atmosphere is related to the fact that the adiabatic lapse rate at large pressures is larger than at low pressures (where, if the air becomes saturated at some height above the surface, the temperature during an adiabatic uplift decreases slowlier than near the surface because of the latent heat released during the condensation process). Conditional instability in the ocean is related to the fact that the adiabatic lapse rate at large pressures is larger than at low pressures

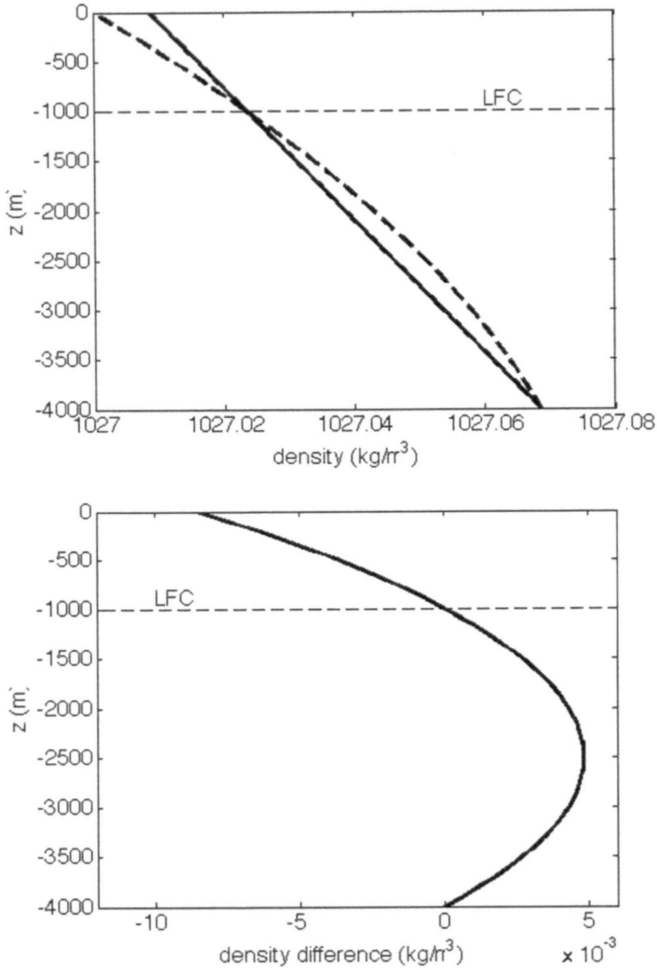

Figure 2. Top panel: Environmental density for a statically stable ocean (solid line), with constant density lapse rate of -0.015 kg m³ km^{-1}, and density that bottom water at temperature 2°C and salinity 35 psu would have if adiabatically brought to level z (dashed line). Bottom panel: difference between the two curves of the top panel. Positive values indicate that bottom water lifted at the given level would be denser than environmental air. Labels indicate the level of free convection (LFC). The simplified equation of state eq.5, and the lapse rate eq.17 have been used to draw this figure. For the sake of a better visualization, here the value of β_p has been reduced to 1% of realistic values. Parameter values: $p_0 = 1$ dbar, $\rho_0 = 1027$ kg/m^3, $T_0 = 2^oC$, $S_0 = 35$ psu, $c_{pl} = 3986$ J kg^{-1} K^{-1}, $\beta_T^0 = 1.67 \cdot 10^{-4}$ K^{-1}, $\beta_s^0 = 7.8 \cdot 10^{-4}$ psu^{-1}, $\beta_p^0 = 4.39 \cdot 10^{-12}$ m s^2 kg^{-1}, $\gamma^0 = 1.1 \cdot 10^{-8}$ Pa^{-1}.

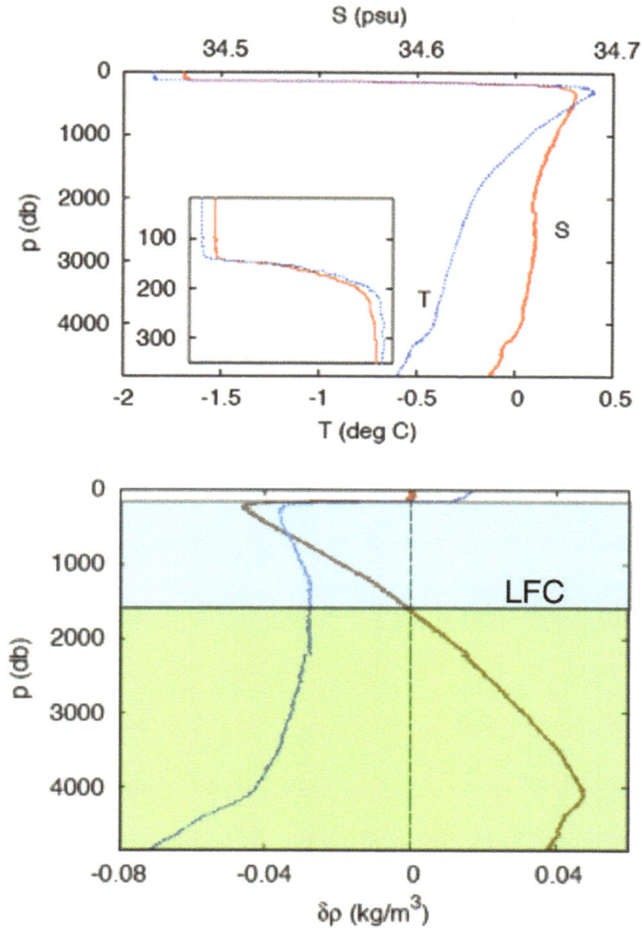

Figure 3. The top panel shows hydrographic data from the Weddell Sea recorded on 29 September 1989 (WOCE dataset). The bottom panel shows the density difference between the density that mixed layer water would have when adiabatically displaced to pressure p and in situ density at pressure p (red line). At about 1700 db (LFC) the difference becomes positive, indicating that surface water becomes denser than environmental water below the level of free convection. The blue line shows the same density difference, but for water from the pycnocline, at 180 db. The full UNESCO equation of state IES80 has been used to draw this figure.

because of thermobaricity: The thermal expansion coefficient for water, for the temperature and pressure ranges relevant for the ocean and unlike most other liquids, increases with pressure (i.e. at large pressures the same temperature change leads to a larger change in density).

3.3 Convective available potential energy

Measures of conditional instability are typically introduced in terms of *convective available potential energy*, CAPE. To define it, we start by writing density as

$$\rho(x, y, z, t) = \rho_0 + \delta\rho(x, y, z, t)$$

where ρ_0 is a reference density and $\delta\rho$ the departure from it. We shall assume $\delta\rho(x, y, z, t) \ll \rho_0$. We also define a reference pressure profile $p_0(z)$ in hydrostatic balance with the reference density $\rho_0(z)$, so that $\frac{dp_0}{dz} = -g\rho_0$ and write

$$p = p_0(z) + \delta p(x, y, z, t).$$

The momentum equation, without any approximation, is then

$$(\rho_0 + \delta\rho)\left(\frac{D\mathbf{u}}{Dt} + 2\mathbf{\Omega} \times \mathbf{u}\right) = -\nabla\delta p - \frac{dp_0}{dz}\mathbf{k} + \rho_0\mathbf{g} + \delta\rho\mathbf{g} + \mu\nabla^2\mathbf{u}$$

Making use of the assumption that $\delta\rho \ll \rho_0$ we get

$$\left(\frac{D\mathbf{u}}{Dt} + 2\mathbf{\Omega} \times \mathbf{u}\right) = -\frac{\nabla\delta p}{\rho_0} + \frac{\delta\rho}{\rho_0}\mathbf{g} + \frac{\mu}{\rho_0}\nabla^2\mathbf{u} \qquad (20)$$

which is the momentum equation in the Boussinesq approximation. The term $b = -\frac{\delta\rho}{\rho_0}g$ is called *buoyancy*.
The vertical component of equation 20 in steady state, neglecting pressure, viscosity and lateral motion, leads to

$$w\frac{\partial w}{\partial z} = -\frac{\delta\rho}{\rho_0}g = b.$$

Integration of this equation yealds the vertical component of the specific kinetic energy:

$$\int w\frac{\partial w}{\partial z}dz = \frac{w^2}{2} = \int b(z)\, dz.$$

If we now consider the term $b(z)$ as the buoyancy that a fluid parcel would have if adiabatically displaced to the level z, we can define the *convective available potential energy*, which measures the work that the atmosphere or the ocean can do on buoyant parcels, at the expenses of internal energy, as:

$$\text{CAPE} = \frac{w^2}{2} = \int_{z_{lfc}}^{z_{blc}} w\frac{\partial w}{\partial z}dz = \int_{z_{lfc}}^{z_{lnb}} bdz,$$

The integration interval is defined by the level of free convection, where convective vertical velocity is zero, and the bounding level of convection, defined as the level at which the parcel becomes neutrally stable with respect to the environmental profile or the physical boundary of the domain. In the atmosphere, such level is called level of neutral stability and is often just above the tropopause, where the density decreases significantly reducing the buoyancy of the displaced parcel. In the ocean in some cases neutral stability is not reached, in which case and for surface driven convection the integration will be taken from the level of free convection to the seafloor. Ingersoll (2005) has calculated the potential energy that can be stored and released by thermobaric instability in the ocean, and Adkins et al. (2005) have discussed possible climatic effects.

4 Discussion and conclusions

Considering that viscosity, pressure gradients and mixing have been neglected in the above derivation, CAPE is no more than an upper limit to the kinetic energy that a fluid parcel can achieve when displaced from rest at height z_0 to a given height z. As such, $w_{max} = 2\sqrt{\text{CAPE}}$ is the maximum vertical velocity a buoyant parcel can achieve. In reality, this upper limit is seldom reached. Clearly, there can be no convection even in presence of large values of CAPE (the above definition of CAPE does not take into account the presence of a barrier to convection that has to be overcome to reach the level of free convection, this barrier is sometimes called CIN, convective inhibition). Furthermore, entrainment of environmental air/water, non hydrostatic adverse pressure gradient forces arising from the vertical displacement of the fluid and viscosity inhibit the vertical motion.

For those reasons, the mere knowledge of convective available potential energy and of its tendency does not provide a real metric of the intensity of convection. For instance, numerical simulations of tropical cyclones have shown that the intensity of hurricanes does not correlate with the environmental CAPE (Shen et al. (2000); Persing and Montgomery (2005)). Even in simple systems, it is not well understood what the efficiency of the conversion of buoyancy fluxes into kinetic energy of the convective motion depends upon.

The most studied set up for convection is the classical Rayleigh-Bénard problem: the Boussinesq approximation for the momentum equation, a linear equation of state for a fluid with constant chemical composition, the advection-diffusion equation for temperature in a divergenceless flow, taken with no slip and fixed temperature boundary conditions at top and bottom boundaries. In such set up, convection develops provided the bottom-top temperature difference is larger than a threshold depending on the fluid viscosity and diffusivity, so clearly the criterion for the development of convective motion is more stringent than the static stability criterion. When this happens, thin unstable boundary layers develop near the boundaries and connect the marginally stable profile in the core of the domain to the boundary conditions. In the unstable boundary layers there is generation of available potential energy, injected into the system through the boundaries. Such energy is transformed by buoyancy fluxes into kinetic energy of the convective motion (eventually balanced by viscous dissipation) and into irreversible mixing. Hughes et al. (2013) showed that, due to the development of large density gradients on small scales, at least 50% of the available potential energy supply is used for mixing of the flow, leaving a smaller fraction directly available for convection.

The system has been studied over the years in many different configurations, to make it more relevant for geophysical fluid dynamics. Considering that tropical convection is often viewed as a quasi-equilibrium process, in which convective clouds consume CAPE at the same rate it is supplied by large-scale processes (Emanuel et al. (1994)), and that radiative-convective equilibrium is the simplest form of such equilibrium, radiative cooling has soon been added to the problem (Prandtl (1925)). Later, the effects of an adiabatic lapse rate have been introduced (Berlengiero et al. (2012)). Complexity has also been increased from the simple Boussinesq model to full three dimensional cloud resolving numerical simulations, and the effects of detailed microphysics and precipitation have been investigated. For instance, it has been shown that the maximum updraft velocity in the convective motion does not merely depend on the buoyancy flux but it is affected by microphysical aspects, such as the size of raindrops (Parodi and Emanuel (2009)). In the last decade, a lot of effort has been devoted to understanding the clustering of the convective towers into bigger structures and how this clustering affects the energy balance and the atmospheric profiles of temperature and moisture (Bretherton et al. (2005); Nolan et al. (2007); Muller and Held (2012); Muller and Bony (2015)).

This short discussion highlights that the prediction of the strength and structure of atmospheric and oceanic convection, and the parameterization of its effects in climate models, are still very active areas of research.

Bibliography

J. F. Adkins, A. P. Ingersoll, and C. Pasquero. Rapid climate change and conditional instability of the glacial deep ocean from the thermobaric effect and geothermal heating. *Quaternary science reviews*, 24:581–594, 2005.

K. Akitomo. Open-ocean deep convection due to thermobaricity: 1. scaling argument. *Journal of Geophysical Research: Oceans*, 104(C3):5225–5234, 1999a.

K. Akitomo. Open-ocean deep convection due to thermobaricity: 2. numerical experiments. *Journal of Geophysical Research: Oceans*, 104(C3): 5235–5249, 1999b.

M. Berlengiero, K. A. Emanuel, J. von Hardenberg, A. Provenzale, and E. A. Spiegel. Internally cooled convection: A fillip for philip. *Commun Nonlinear Sci Numer Simulat*, 17:1998–2007, 2012.

C. F. Bohren and B. A. Albrecht. *Atmospheric thermodynamics.* Oxford University Press, 2009.

C. S. Bretherton, P. N. Blossey, and M. Khairoutdinov. An energy-balance analysis of deep convective self-aggregation above uniform sst. *Journal of Atmospheric Sciences*, 62:4273–4292, 2005.

J. A. Curry and P. J. Webster. *Thermodynamics of Atmospheres and Oceans.* Academic Press, 1999.

K. A. Emanuel. *Atmospheric convection.* Oxford University Press, 1994.

K. A. Emanuel, D. J. Neelin, and C. S. Bretherton. On large-scale circulations in convecting atmospheres. *Quarterly Journal of the Royal Meteorological Society*, 120(519):1111–1143, 1994.

N. P. Fofonoff. Physical properties of seawater: A new salinity scale and equation of state of seawater. *Journal of the Geophysical Research*, 90 (C2):3332–3342, 1995.

A. E. Gill. *Atmosphere-Ocean dynamics.* Academic Press, 1982.

G. O. Hughes, B. Gayen, and R. W. Griffiths. Available potential energy in Rayleigh-Bénard convection. *Journal of Fluid Mechanics*, 729, 8 2013. ISSN 1469-7645. doi: 10.1017/jfm.2013.353.

A. P. Ingersoll. Boussinesq and anelastic approximations revisited: potential energy release during thermobaric instability. *Journal of Physical Oceanography*, 35:1359–1369, 2005.

M. G. McPhee. Is thermobaricity a major factor in southern ocean ventilation? *Antarctic Science*, 15:153–160, 3 2003. ISSN 1365-2079.

C. J. Muller and S. Bony. What favors convective aggregation, and why? *Geophysical Research Letters*, 42:5626–5634, 2015.

C. J. Muller and I. M. Held. Detailed investigation of the self-aggregation of convection in cloud-resolving simulations. *Journal of Atmospheric Sciences*, 69:2551–2565, 2012.

D. S. Nolan, E. D. Rappin, and K. A. Emanuel. Tropical cyclogenesis sensitivity to environmental parameters in radiativeconvective equilibrium. *Quarterly Journal of the Royal Meteorological Society*, 133(629):2085–2107, 2007.

A. Parodi and K. Emanuel. A theory for buoyancy and velocity scales in deep moist convection. *Journal of Atmospheric Sciences*, 66:3449–3463, 2009.

J. Persing and M. T. Montgomery. Is environmental cape important in the determination of maximum possible hurricane intensity? *Journal of the Astmospheric Sciences*, 62:542–550, 2005.

L. Prandtl. Bericht über untersuchungen zur ausgebildeten turbulenz. *Zs .angew. Math. Mech.*, 5:136–139, 1925.

K. Schwarzschild. On the equilibrium of the sun's atmosphere. *Nach. Gesell. Wiss. Göttingen*, 195:41–53, 1906.

W. Shen, R. E. Tuleya, and I. Ginis. A sensitivity study of the thermodynamic environmental of gfd model hurricane intensity: Implications for global warming. *Journal of Climate*, 13:109–121, 2000.

B. Stevens. *Class notes on atmospheric moist convection.* URL `http://www.mpimet.mpg.de/fileadmin/staff/stevensbjorn/teaching/Thermodynamic_Notes.pdf`. Retrieved on March 31st, 2015.

B. Stevens. Atmospheric moist convection. *Annual Review of Earth and Planetary Sciences*, 33:605–643, 2005.

Z. Su, A. P. Ingersoll, and A. F. Thompson. Convective Available Potential Energy of World Ocean. *AGU Fall Meeting Abstracts*, page C1749, December 2012.

G. Vallis. *Atmospheric and oceanic fluid dynamics.* Cambridge University Press, 2006.

Laboratory experiments on large-scale geophysical flows

Miklós Vincze [*‡] and Imre M. Jánosi [†‡]
[*] MTA-ELTE Theoretical Physics Research Group, Budapest, Hungary
[†] Department of Physics of Complex Systems, Eötvös Loránd University, Budapest, Hungary
[‡] von Kármán Laboratory for Environmental Flows, Eötvös Loránd University, Budapest, Hungary

1 Historical Overview

Laboratory experiments have been playing a unique role in the great scientific endeavor to better understand the complex phenomena of environmental flows for centuries. Thus, we find it quite appropriate to start our review – motivated by an Italian conference and written by Hungarian authors – with a historic reference to the heritage of the 17th century scholar Luigi Fernando *Marsigli* (1658–1730), native of Bologna. Being eminent in all areas of contemporary science ranging from mathematics to anatomy, he was commissioned to carry out the complete cartographic and zoological survey of the border area between the Kingdom of Hungary and the Ottoman Empire: a task that made him spend twenty years of his life roaming in and around Hungary. His visit to Constantinople (now Istanbul, Turkey) inspired his later pioneering work on the physical background of seawater exchange through the strait of Bosporus. He assumed that the reported undercurrent from the Mediterranean towards the Black Sea is driven by the density difference between the two basins due to the higher salinity of Mediterranean seawater. Marsigli demonstrated this idea in an experimental tank which he divided into two sections by a vertical wall in the middle (Figure 1b). The two sections were filled with saline- and freshwater and one side was painted with ink for visualization (Marsigli, 1725). The flow was then initiated by opening two holes at the top and bottom of the separating wall, leading to the formation of a saline front at the bottom and an opposing freshwater front at the top. Marsigli's hypothesis (and its demonstration) has led to the theory of buoyancy-driven environmental flows and eventually to the foundation of physical oceanography. The two-box experimental configuration itself has also become one of the standard

A. Provenzale et al. (Eds.), *The Fluid Dynamics of Climate*,
CISM International Centre for Mechanical Sciences
DOI 10.1007/ 978-3-7091-1893-1_3 © CISM Udine 2016

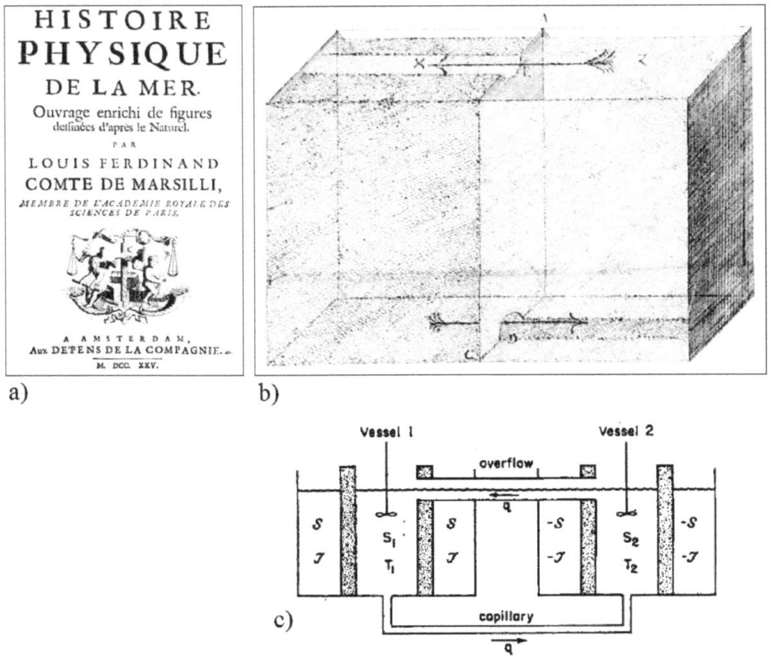

Figure 1. (a) The cover of probably the very first book on physical oceanography, Marsigli's "Histoire physique de la mer", translated to French by Leclerc (Amsterdam, 1725). (b) The schematic drawing of Marsigli's two-box experimental set-up from the book. (c) The sketch of Henry Stommel's famous conceptual model of the meridional overturning from his original paper (Stommel, 1961).

conceptual tools in the field to this day: the theoretical study of the fluxes in a similar set-up with additional differential heating provided the first qualitatively correct insight into the dynamics of the meridional overturning ocean circulation in the groundbreaking works (e.g., (Stommel, 1961)) of Henry *Stommel* (1920–1992), see Figure 1c.

Revealing the pathways of heat exchange and thermal mixing in the overturning currents keeps oceanographers busy even nowadays. One of the most cited papers on these issues also happens to be an experimental work, that of Swedish oceanographer Johan *Sandström* (1874–1947) from 1908

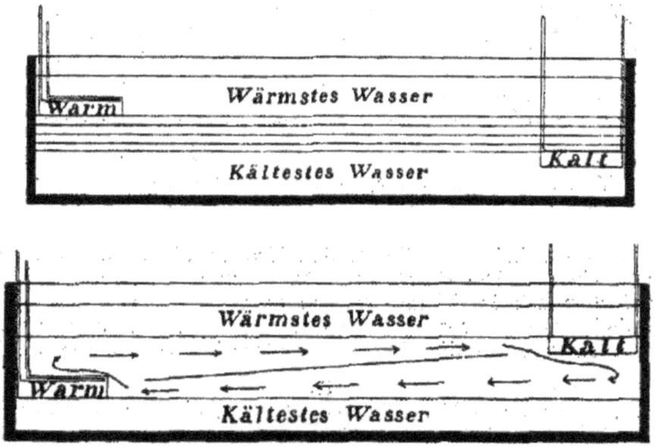

Figure 2. Two sketches from the original paper of J. Sandström on his ocean circulation experiment (Sandström, 1908). According to his 'theorem', if the heat source is at higher level than the heat sink (top) the fluid stays at rest, and overturning can by initiated only if the configuration is reversed (bottom).

(Sandström, 1908). In this study the author investigated how the relative vertical position of a heat source and a heat sink affects the flow properties in a rectangular tank of saline water. He reported that an overturning flow could only be initiated if the heat source was placed *below* the level of the cooling source (sink), as shown in Figure 2, reproduced from the original paper. This conclusion has become known as Sandström's 'theorem' and has been widely disputed in the past century, due to its controversial oceanographic implications. Clearly, the Atlantic meridional overturning circulation does exist, despite being driven by the differential incoming Solar heat fluxes between the equatorial and polar regions, which both act at the water surface, i.e. at the same geopotential level, virtually violating the 'theorem'. Although a recent study has found Sandström's results inconsistent with those of a reconstructed experiment (Coman et al., 2006), yet the work's impact is undeniable: to circumvent the violation of the 'theorem', researchers have uncovered various possible ways of heat and momentum transfer in the ocean, from localized geothermal heat sources (Vincze et al., 2011b) to tidally excited internal waves (Zhang et al., 2008).

The first experimental demonstration of such large-amplitude internal waves that propagate in the bulk of the stratified ocean while remaining

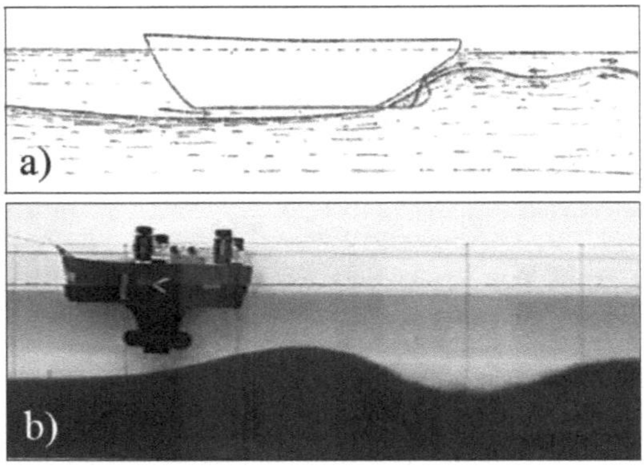

Figure 3. (a) A schematic drawing of internal wave excitation behind a ship model from Ekman's original paper (Ekman, 1904). (b) The experiment repeated a hundred years later in the von Karman Laboratory, Budapest.

practically unnoticed from the surface has been conducted by Sandström's fellow citizen and contemporary, Vagn Walfrid *Ekman* (1874–1954). Ekman was given captain Nansen's logs on the 1893-96 Arctic expedition of the legendary research ship Fram. Nansen wrote: "When caught in dead water Fram appeared to be held back, as if by some mysterious force, and she did not always answer the helm. In calm weather, with a light cargo, Fram was capable of 6 to 7 knots. When in dead water she was unable to make 1.5 knots." (Walker, 1991). By towing a model of the ship in a laboratory tank filled with stratified water (Figure 3), Ekman realized that this motion excites large waves in the bulk at the internal density interfaces, to which the model ship loses its kinetic energy, hence, speed (Ekman, 1904). This demonstration paved the way for the research of flow-topography interactions in stratified media. It is also interesting to note that Nansen's very same notebook contained the descriptions of the drift of icebergs as well, which motivated Ekman's discovery of the Ekman spiral, a contribution that has clearly made him one of the founding fathers of modern oceanography and meteorology.

Besides stratification, the other important unique feature that distinguishes geophysical hydrodynamics from the everyday 'human-scale' fluid mechanics is the dominant role of Coriolis force. A surprisingly large part of the complexity of atmospheric and ocean dynamics at the mid-latitudes can be attributed to the interplay of the planetary rotation (yielding Coriolis effect) and the horizontal buoyancy (temperature and/or salinity) gradients. The first experimental 'minimal model' of atmospheric circulation based on these principles was constructed by German meteorologist and medic Friedrich *Vettin* (1820–1905) in 1857. He studied the flow patterns of cigar smoke in an air-filled rotating cylindrical container, in which radial temperature gradient was introduced by a block of ice placed at its center (Vettin, 1884), representing the cold polar regions of Earth. Vettin observed convective vortices, but the applied model parameters (size, rotation rate, temperature gradient) did not allow the exploration of flow regimes outside the 'sideways-convective' axially symmetric overturning state (also known as Hadley regime).

These pioneering studies and the similar ones that followed in the first half of the 20th century (Fultz, 1949, 1952) of rotating convective flows were rather qualitative in nature and their experimental circumstances and procedures were hard to control and replicate. Yet, the same period brought forth the first systematic laboratory studies which could be *quantitatively* compared with linear stability analyses in the field of rotating fluids, namely the classic works of Sir Geoffrey Ingram *Taylor* (1886–1975) on the cylindrical Couette flow (Taylor, 1923). It was not until the 1950s when the differentially heated and rotating configuration was investigated to such detail and precision by the team of David *Fultz* (1921–2002) in Chicago and, in parallel, the young Raymond *Hide*, then graduate student in Cambridge (UK). In his PhD thesis, Hide gave the first description of periodic waves, analogous to the large-scale planetary waves in the atmosphere, in an annular rotating tank filled with water, cooled at the inner cylindrical sidewall and heated at the outer rim (see Figure 4). As remarked by Ghil et al. (2010), the thesis also contained "one of the first, or maybe *the first*, study of what we call today bifurcation and regime diagrams in a fluid dynamical context". The insight provided by these experiments into the basic underlying dynamics of atmospheric circulation had a very powerful impact on the formulation of theoretical and numerical minimal models (conceptual, or 'toy models') of the atmosphere, i.e. the ones constructed by Edward Norton *Lorenz* (1917–2008), see e.g. Lorenz (1963); the experiments have clearly demonstrated which factors are truly essential for an 'irreducible' description of the basics of mid-latitude atmospheric flows, and which ones can be omitted. It turned out that for a minimal model of cyclogenesis or jet stream formation one does not need to consider the role of topography, evaporation, precipitation or even the curvature of the surface, only the meridional temperature difference and the Coriolis effect (Lorenz, 1967).

This approach of creating environmentally inspired conceptual laboratory models has been flourishing since then: currently active research topics range from modeling the interactions between internal waves and various seafloor topographies (Zhang et al., 2008; Boschan et al., 2012) to the experimental exploration of "meddy" formation in rotating stratified media (Aubert et al., 2012), wave attractors of inertial and gravity waves (Klein et al., 2014) or the conceptual demonstration of the famous hexagonal vortex at the north pole of planet Saturn, just to pick a random selection, in which all studies share the main principle of the above listed historic experimental works: making the model of the given phenomenon "as simple as possible, but no simpler".

Inspired by the fascinating property of nature that planet-scale hydrodynamic phenomena can be properly modeled using tabletop-size setups in a relatively inexpensive way, and the perspective that even physical oceanographic problems can be addressed in a completely land-locked country, triggered the foundation of the *von Karman Laboratory of Environmental Flows* at the then-new campus of the Eötvös University (Budapest, Hungary) in 2000. Of course, in terms of instrumentation, the aim of our laboratory has never been to compete with the impressive capabilities of the handful of major institutions in Europe (e.g. the Geophysical Fluid Dynamics Lab of the Department of Physics in Oxford, or the 13 m-wide Coriolis Platform in Grenoble, where all measurement devices and even the experimenters can co-rotate with their set-ups). Nevertheless, our experience of the past decade has clearly shown that there is a lot to discover in this – surprisingly young – field of classical physics, where even the simplest experiments can yield useful insight into the underlying dynamics of environmental flows, if one looks carefully enough. In the present review, after giving a brief introduction into the principles that make such modeling possible, we present a few 'case studies' selected among our more recent results to demonstrate the diversity and efficiency of this experimental approach and the ways it can contribute to our better understanding of the climate system.

2 The basics of laboratory modeling

For a brief mathematical demonstration of the principle of *hydrodynamic similarity*, let us consider the Navier-Stokes equation of an incompressible flow, as observed from a rotating system:

$$\frac{d\vec{v}}{dt} = -2\vec{\Omega} \times \vec{v} - \frac{1}{\rho}\nabla p - \vec{n}g + \nu\nabla^2\vec{v}. \tag{1}$$

Here $\vec{\Omega} \equiv \Omega\vec{n}$ denotes the angular velocity vector of the reference frame (co-aligned with the vertical unit vector $\vec{n}$), $\vec{v}$ is the velocity vector, ρ and ν represent the density and kinematic viscosity of the fluid parcel, respectively,

g represents the gravitational acceleration and p is the pressure field. The form $d \cdot /dt \equiv \partial \cdot /\partial t + \vec{v}(\nabla \cdot)$ on the left hand side denotes the advective time derivative.

One can now rescale the equation by setting the linear size L of the considered domain, and the flow's characteristic (e.g., mean) velocity U as 'natural' units. These transformations can be expressed formally as replacements $\vec{r} \to L\vec{r}$ ($\vec{r}$ being the position vector) and $\vec{v} \to U\vec{v}$. Combining the new length and velocity units yields our natural timescale: $t \to L/U\, t$. Though not that manifestly, but one can easily get the corresponding transformation of the pressure units as well (for details we refer to, e.g. Vallis (2006)) in the form of $p \to 2\rho|\Omega|LU\, p$. After carrying out these replacements and dividing the equation by the scale factor U^2/L of the rescaled derivative at the left-hand side we arrive at the following re-organized form:

$$\frac{d\vec{v}}{dt} = -\frac{1}{Ro}(\pm\vec{n} \times \vec{v} + \nabla p) - \frac{1}{Fr^2}\vec{n} + \frac{1}{Re}\nabla^2\vec{v}, \tag{2}$$

where Rossby number $Ro \equiv U/(2|\Omega|L)$ quantifies the characteristic ratio of hydrodynamic acceleration and Coriolis acceleration (the appearance of $\pm$ is a consequence of defining Ro with the absolute value of the angular velocity). Similarly, Froude number $Fr \equiv U/\sqrt{gL}$ accounts for the relative magnitude of gravitational acceleration, and Reynolds number $Re \equiv UL/\nu$ measures the importance of the viscous acceleration in the same manner. It is to be emphasized that these products are dimensionless pure numbers.

Two flows are considered dynamically similar if – besides the geometric similarity of the set-ups – their dimensionless numbers match. Note, that it is generally not possible to fulfill perfect similarity in terms of *all* parameters, unless the two considered systems are identical. However, in most of the applications many terms can be neglected. For the vast majority of engineering-oriented applications the Coriolis force does not play any role ($Ro \to \infty$), and the effect of gravity (hydrostatic pressure) can be absorbed into the pressure term. Thus, for a wind-tunnel test of a car model, Reynolds number Re is the only relevant, and perfectly sufficient, parameter of dynamic similarity. For a ship model in an open channel flow, however, both viscous drag (quantified via Re) and gravity wave emission at the water surface (scaling with Fr) dissipate some of the model's kinetic energy. In this case both forces are of relevance, and the full similarity could only be achieved (if at all) by using fluids of different viscosities in the model and the prototype (Kundu and Cohen, 2008).

Fortunately, changing the medium is not the only possible way to circumvent such difficulties. Laboratory experiments can still be useful even without reaching perfect similarity, if one is able to separate the differently scaling effects and apply corrections accordingly. For example, in the above case, the viscous drag acting on the ship model can be measured alone in a control experiment, where the surface wave excitation is somehow inhibited (e.g. by placing a rigid lid at the top of the tank), and therefore only the value of Re remains relevant. By subtracting this result from the drag of the original experiment, one can separate the contribution of the emitted surface waves, and this difference can be readily compared to the large prototype.

For planet-scale flows, the first term of (2) predominates ($Ro \to 0$). The stationary solution of the governing equation (2) in this limit can be obtained by omitting the gravitational and viscous terms ($Fr \to \infty$ and $Re \to \infty$) and setting the left hand side to zero. In this case, referred to as *geostrophic equilibrium*, pressure perturbations are balanced by stationary flow *along* the isobars. In environmental flows, generally, the larger the considered length scale is, the better this approximation becomes, as the characteristic values of Ro drop. For example, in case of a cyclone with a linear size $L = 1000$ km and wind speed $U = 10$ m/s the formula yields $Ro \approx 0.07$, and can be even two or three orders of magnitude smaller for meddies (i.e. isolated vortex lenses in the salty Mediterranean water). As a comparison, for a bathtub drain (with $U = 0.1$ m/s and $L = 0.1$ m), one gets ca. $Ro \approx 6850$. Thus, by taking a quick look at (2), it is immediately visible, that the widespread rumors about bath drain flows circulating opposite ways on the Northern and Southern hemispheres due to Coriolis effect have no basis at all: besides the characteristic values of Fr and Re of the same flow, this effect is completely negligible. However, by placing the same bathtub flow onto a turntable with an angular velocity $\Omega = 7$ rad/s (about one rotation per second), one can arrive at the 'cyclonic' value of Ro in laboratory scale, making it relatively easy to achieve hydrodynamic similarity in a rotating model flow.

Obviously, the governing equation (2) is not sufficient for a complete physical description of the flow in itself. In case of a large-scale environmental flow, the thermal boundary conditions and the 'equation of state' – here, the relationship $\rho(T)$ between the density and temperature of the fluid parcels – are of fundamental importance just as well. A convenient nondimensional combination for quantifying all these factors in a rotating thermal convection driven by a horizontal temperature gradient is a special version of Ro. It is known as thermal Rossby number Ro_T (or Hide number), and defined as

$$\mathrm{Ro}_T = \frac{\alpha g H \Delta T}{(2\Omega)^2 L^2}\ , \tag{3}$$

where α is the coefficient of volumetric thermal expansion for the fluid and ΔT is the temperature contrast imposed at the vertical boundaries of the rotating layer.

The thermal Rossby number has a fairly transparent explanation by considering the aforementioned case of geostrophic equilibrium. Expressing the relevant terms of (2) in cylindrical coordinates, the zonal velocity component u_θ is determined by the radial pressure difference as

$$u_\theta = -\frac{1}{\rho_0 2\Omega}\frac{\partial p}{\partial r} \ , \tag{4}$$

where ρ_0 is a mean density of the fluid at a reference temperature T_0, and friction is neglected. The pressure difference is a consequence of radial temperature contrast inducing a change in density $\Delta\rho = -\alpha\rho_0\Delta T$. Using the hydrostatic approximation $p = \rho g H$, the radial pressure difference in a shallow layer can be estimated as $\Delta p = -\alpha\rho_0 g H\Delta T$, which gives an estimate to a relative velocity scale

$$U_T = \frac{\alpha g H \Delta T}{2|\Omega|L} \ . \tag{5}$$

A simple comparison with Eq. (3) reveals that the thermal Rossby number is in full analogy with the "regular" one, since it is given as the ratio of two characteristic velocity scales:

$$\mathrm{Ro}_T = \frac{U_T}{2|\Omega|L} \ . \tag{6}$$

We note, that when the horizontal temperature difference ΔT is comparable to the vertical one, i.e. $\Delta T_z \approx \Delta T$ holds, Ro_T is proportional to the Burger number Bu, defined as the squared ratio of the Rossby deformation radius $R_d = \sqrt{gH\alpha\Delta T_z}/\Omega$ to the linear size L; $Bu = gH\alpha\Delta T_z/(\Omega^2 L^2)$.

As we will soon see, Ro_T is not the only relevant parameter to characterize thermally driven rotating flows, yet, it clearly demonstrates the power of hydrodynamic similarity: by adjusting the thermal boundary conditions, geometric properties and the rotation rate in a model setup, it is possible to access various spatial and temporal scales of the atmospheric and oceanic flows, even to model atmospheric features of other planets in the solar system, see e.g. Mitchell et al. (2015).

3 Case studies

In the section that follows four current or recent experimental research topics of the authors will be introduced and briefly discussed, all of which are

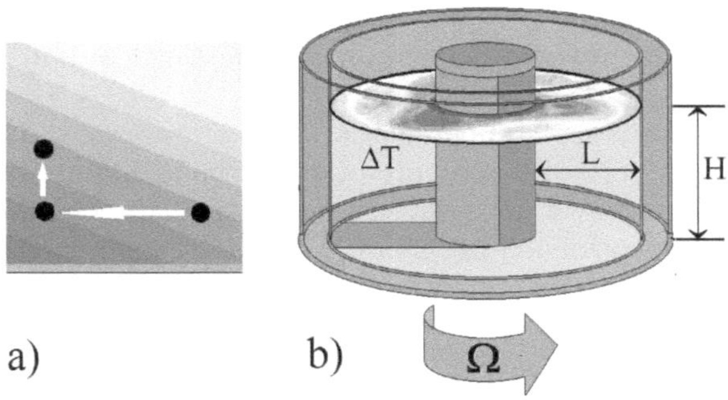

Figure 4. (a) A conceptual sketch of baroclinic unstable fluid parcel displacement through sloping isothermal levels. (b) A schematic drawing of a baroclinic wavetank with the most important geometrical and physical parameters.

related to problems of thermally driven rotating environmental flows. We selected these case studies so that they can hopefully give the reader an impression of the diversity of phenomena that can be studied by using slightly different variants of practically the same experimental configuration.

3.1 Baroclinic instability and inertia-gravity waves

Without the effect of rotation, the constant density levels in a stably stratified system would be perfectly horizontal. In geostrophic equilibrium, however, winds blowing perpendicular to the plane of the local density gradient can stabilize tilted stratification – a direct consequence of equation (4). This process can also act the other way: if constant heat fluxes at the vertical boundaries maintain a sloping a stratification in a rotating system, an out-of-plane meanflow, the so-called *thermal wind* will develop. Within the framework of the geostrophic theory, one can easily obtain a connection between windspeed U and the slope s_T of isotherms (see, e.g. Vallis (2006)) in the form of

$$s_T = \frac{2|\Omega|U}{g\alpha\Delta T}. \tag{7}$$

If s_T is large enough, a fluid parcel, displaced by an initial perturbation parallel to the (flat) bottom boundary, though traveling along an equipotential, may reach a region of higher density (Figure 4a). From here, the buoyancy force lifts the parcel until it reaches its original isotherm – at a *larger* geopotential. Apparently, a small perturbation yields a larger displacement, thus this flow is unstable; the phenomenon is hence coined *baroclinic instability*. This 'surplus' potential energy is released in the form of kinetic energy, and excites baroclinic (planetary) waves, the main sources of cyclogenesis in the mid-latitude atmosphere. Besides Ro_T the kinematic viscosity ν of the medium also plays an important role in this instability; it introduces a 'viscous cutoff' that dissipates too weak thermal winds and also damps the instability of too large wavenumbers. This effect is parametrized by Taylor number Ta that accounts for the ratio of rotational and viscous effects, and reads as

$$Ta = \frac{4\Omega^2 L^5}{\nu^2 H}. \tag{8}$$

Ro_T and Ta are thus used together to characterize the different dynamical regimes in rotating, thermally driven systems, such as planetary atmospheres, oceanic basins and their minimal models in the laboratory. A typical experimental configuration is depicted in Fig.4b: the tank consists of co-rotating coaxial cylinders. The innermost cylinder (representing the polar region) is cooled whereas the outer rim is heated, thus the working fluid in the middle annular gap experiences temperature difference ΔT. Traditionally, for this experimental set-up the width of the gap (i.e. the difference of its inner and outer radii) is used as characteristic length L.

In Figure 5 we present a sketch of the regime diagram (after Hide (1953)) defined by the two nondimensional numbers. The insets (labelled with letters a-e) show examples of the typical flow patterns of the different regimes. The Ta and Ro_T values corresponding to the insets are highlighted in the regime diagram. Figure 5a is a composite satellite image of Earth's southern hemisphere as seen from the direction of its axis (with Antarctica in the middle). The irregularly shaped cyclonic and anticyclonic eddies of Earth's atmosphere are visibly quite similar in structure to those obtained via infrared thermography of the rotating annulus at the Brandenburg Technical University (BTU) Cottbus, Germany, at similar values of Ta and Ro_T (Figure 5b). Decreasing Ta and increasing Ro_T simultaneously by two orders of magnitude brings the system to the regime where the velocity and temperature fields exhibit persistent regular waves that propagate along the azimuthal direction in the tank due to baroclinic instability. Figure 5c shows a typical snapshot of such a pattern with dominant wavenumber (i.e. the number of lobes) $m = 3$ in the annulus. Figure 5d and e show the axisymmetric structures that develop when further decreasing Ta and increasing Ro_T. The former one is, again, a thermographic snapshot of the annulus, whereas the latter is a spacecraft image of the atmosphere of planet Venus, well known for its slow rotation (one revolution takes 243 Earth days). It

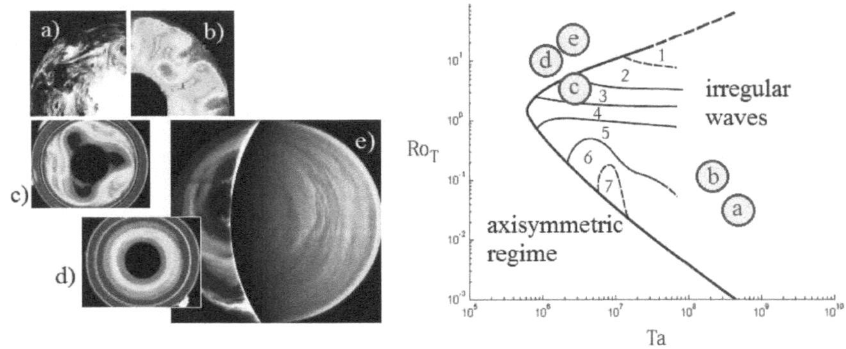

Figure 5. The schematic regime diagram of thermally driven rotating flows, defined by Taylor and Rossby numbers. The numbers indicate the azimuthal wavenumbers of baroclinic waves along the circumference of the tank for the 'regular wave' regime. The insets: the mid-latitude atmosphere of Earth (a), thermographic images of the BTU rotating annulus setup at the different flow regimes (b-d) and a composite image of Venus' atmosphere; the sun-lit side in visible, the night side in infrared wavelength range (Source: ESA).

can be generally stated that for large values of Ta (far from the 'viscous cutoff'), in case of $Ro_T \gg 1$, the flow is axisymmetric and not significantly disturbed by rotation (as in the Venusian atmosphere), whereas for $Ro_T \ll 1$ the dynamics is dominated by the Coriolis effect.

As briefly mentioned in the historical overview, internal (gravity) waves play an important role in heat and momentum transport both in the oceans and the atmosphere. In a stably stratified system, a fluid element with reference density ρ_0 displaced vertically from height z, keeping its density, oscillates at buoyancy- (or Brunt–Väisälä-) frequency $N(z) = \sqrt{g\rho_0^{-1}\,|d\rho(z)/dz|}$. Gravity waves can be understood as propagating buoyancy oscillations in the medium. These 'ripples' are widely known to be generated by flow-topography interactions (lee waves in the atmosphere or tidal conversion above mid-ocean ridges). Yet it has been noticed just recently by meteorologists (Ghil et al., 2010) that similar waves are also excited in the vicinities of the large-scale baroclinic cyclones without any topographic obstacle, and they are increasingly recognized as significant sources of uncertainty in weather forecasting. Strictly speaking, these waves are not pure gravity waves, since the Coriolis effect (inertial motion) also affects their propagation. Such "hybrid" combinations of the two wave types are referred to as inertia-gravity waves (IGWs), and their nontrivial dispersion relation has the form of

$$\omega^2 = N^2 \cos^2 \zeta + f^2 \sin^2 \zeta, \tag{9}$$

where the angular frequency ω of the wave is determined by the aforementioned N, the Coriolis frequency $f \equiv 2\Omega \sin \vartheta$ (Ω being the angular frequency of the rotation of Earth and ϑ the geographic latitude), and $\zeta \equiv \arctan(n/\sqrt{k^2 + l^2})$, the angle of phase propagation relative to the horizontal plane, set by horizontal wavenumbers k, l and vertical wavenumber n. Nonlinear interactions between baroclinic waves, spontaneous imbalance and geostrophic adjustment are all possible sources of IGWs in such a system. It is important to note that the scale difference between these waves and the large planetary waves that excite them is very large, making the coupling very difficult to resolve in numerical models. Thus laboratory experiments appear to be useful test-beds to better understand these phenomena, even though perfect hydrodynamical stability can hardly be reached in such a multiscale system.

In the upper troposphere of the mid-latitudes the ratio of N/f is approximately on the order of $\mathcal{O}(100)$, whereas for a typical rotating annulus experiment (in the regime of baroclinic instability) with parameters $L \approx H \approx 0.1$ m, $\Omega \approx 5$ rpm and $\Delta T \approx 10$ K typically $N/f \sim \mathcal{O}(0.1)$ holds (Borchert et al., 2014). Taking a glimpse at equation (9) it becomes obvious that this difference is not just quantitative but also qualitative: in the atmosphere, the high frequency waves propagate nearly horizontally (and the low frequencies vertically), whereas in this classic laboratory set-up the behavior is just the opposite.

There have been several attempts and suggestions to create a more 'atmosphere-like' experimental apparatus, based on the rule of thumb that in such rotating systems where the isothermal surfaces are tilted, the total vertical difference ΔT_z is comparable to its horizontal counterpart ΔT. Decreasing fluid depth H and increasing ΔT both tend to raise the vertical temperature gradient, and thus N, which can be estimated from $N^2 \approx g\alpha\Delta T/H$. However, the water layer cannot be arbitrarily shallow either, because then the boundary layers at the horizontal lids (e.g. the Ekman layers) would invade the flow too much and modify the dynamics markedly. Also, care must be taken not to suppress baroclinic instability, meaning that the geometrical parameters and the boundary conditions must be set so to keep the system within the baroclinic unstable regime of the $Ta - Ro_T$ plane (see Figure 5). For example if f (that is $f = 2\Omega$ for a laboratory tank) were set too small, then the flow would be axially symmetric and no planetary waves would be present to excite IGWs.

In their recent work, Borchert et al. (2014) from the Goethe University of Frankfurt am Main, suggested a wide and shallow configuration ($H = 4$ cm, $L = 50$ cm, $\Omega = 0.78$ rpm, $\Delta T = 30$ K) to yield at least $N > f$ on average. This tank is currently under construction at the Brandenburg Technical University within the framework of the MS-GWaves program, dedicated to the numerical, experimental, and observational investigation of IGWs. In this apparatus the large vertical aspect ratio L/H may lead to difficulties, due to the significant heat exchange occurring between the large horizontal surface of the working flow and the (supposedly insulating) horizontal lids.

Another way to increase N is introducing vertical salinity gradient to the set-up. This method was applied within the framework of a joined German-French-Hungarian collaboration (involving Patrice *Le Gal* of U. Aix-Marseille, Uwe *Harlander* of BTU Cottbus and the authors). A continuously stratified salinity profile was prepared in the experimental cavity before each measurement with the so-called double-bucket technique (Oster and Yamamoto, 1963) two vessels containing saline and fresh water are connected via a pipe so that the freshwater inflow to the saline bucket yields a mixture of ever decreasing salinity in time. By raising or lowering one of the buckets during this filling-up procedure, we were able to adjust the water fluxes into the mixing compartment and, thus, create arbitrarily non-uniform stable salinity profiles.

Imposing lateral temperature and vertical salinity gradients in the same set-up can lead to the formation of *double diffusive convection* (Chen et al., 1971). However, at depths where the salinity stratification is steep enough to make the parameter $\eta \equiv \alpha\Delta T/(d\rho/dz)$ smaller than a critical ratio η^*, the fluid stays at rest: the large gradient inhibits the overturning convection to penetrate the full depth of the bulk, and confines baroclinic instability into a shallower layer, where $\eta > \eta^*$ holds. Therefore the effect of salinity stratification is twofold. Firstly, it directly increases N. Secondly, with reducing the effective H to the vertical domain of the convective layer, it keeps the system baroclinic unstable for smaller values of f as well (note that both Ta and Ro_T scale with Ω^2/H, thus one can trade rotation rate for water depth and still stay in the same flow regime). Moreover, this shallow layer is practically free from the undesirable boundary layer effects, since it can be set 'floating' at any water level, depending on the prepared salinity profile only, without any direct thermal or mechanical connection to the bottom or top lids.

Using this method we were able to reach frequency ratios $N/f \approx 2-6$ in our experiments. These, being larger than unity, imply that the IGW propagation is expected to be qualitatively similar to the atmospheric case. An example of the detected short-wavelength, high-frequency wave patterns is

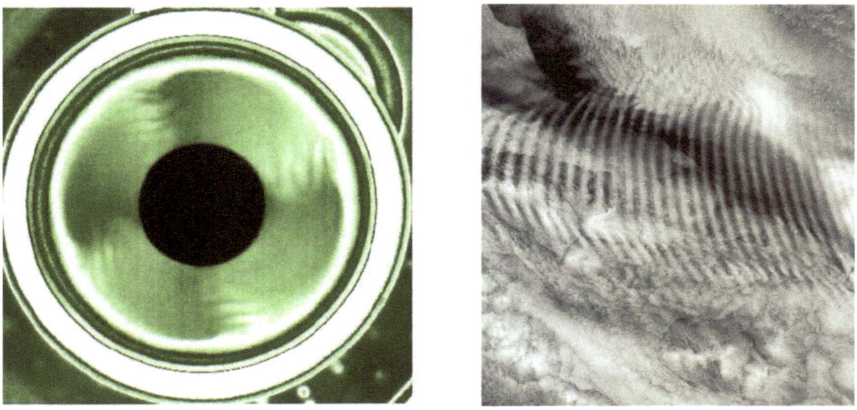

Figure 6. Possible signs of small-scale inertia-gravity wave excitation from the $m = 4$ baroclinic wave pattern (left) and its analogy in the atmosphere (right), as seen from orbit (Source: NASA).

shown in the infrared thermographic image at the left hand side of Figure 6. It is clearly visible that the large-scale fourfold symmetric ($m = 4$) baroclinic waves (anticyclones) radiate away the minor ripples corresponding to the possible surface signatures of IGWs. For a comparison of the structures, an orbital image taken from the International Space Station is presented at the right hand side of Figure 6, showing actual atmospheric IGWs.

The reason of the observed wave emission is probably transient geostrophic imbalance: all the detected IGW-like patterns appeared right after changing rotation rate Ω during the experimental sequence, and vanished within 30 revolutions time, whereas the large-scale baroclinic wave patterns survived throughout the whole measurement period (typically lasting for 2-3 hours or 6-700 revolutions). It is to be remarked that infrared thermography is far not optimal method for IGW detection, since the applied wavelength range can only escape from the uppermost few millimeters below the free water surface, thus everything that could be observed this way are just faint hints of the dynamics in the bulk.

The search for underwater IGW signatures in the rotating annulus will continue with utilizing an array of small digital thermometers and salinity sensors that will be placed into the working fluid, and will log the values with a temporal resolution above 10 Hz. According to numerical results (Borchert et al., 2014; Randriamampianina, 2013) and the above presented preliminary surface temperature data, the wave filaments appear to be connected to the baroclinic wave lobes. Thus, one would expect that if a baroclinic wave lobe passes by the sensor, amplitude enhancement would appear in the IGW frequency range, i.e. at $f < \omega < N$. The temporal pattern of this modulation itself should therefore be determined by the low-frequency baroclinic wave propagation. The signals of different temporal scales will then be separated and compared via bandpass filtering. These laboratory experiments are currently underway, and will hopefully yield clear experimental evidence of IGW excitation processes in a differentially heated rotating set-up.

3.2 Dynamics of passive tracers in the atmosphere

Anthropogenic emissions often result in pollution levels that exceed air quality standards at many locations over both hemispheres. Air quality and pollutant dispersal are strongly influenced by large scale transport processes at the intercontinental and global scales. Pollutant particles can travel huge distances depending on their size and density, small aerosols can reach the stratosphere and get around the globe several times until falling out. The first approximation for the motion of light particles advected by environmental flows is the approach of "passive tracer" dynamics, where particle size is assumed to be negligible and the density is equal with the surrounding liquid, thus such particles passively follow the flow without disturbing it (see the Chapter by Haszpra and Tél in this volume). Passive tracer dynamics is often studied experimentally by using various dyes (Ewart and Bendiner, 1981; Sommerer et al., 1996; Gouillart et al., 2009), since soluble chemicals hardly affect flow fields especially in the turbulent regime.

Our experiments (Jánosi et al., 2010) were carried out in the classical laboratory model for mid-latitude large-scale flow phenomena, which is a differentially heated rotating annulus invented by Fultz (Fultz, 1949, 1952; Fultz et al., 1959) and Hide (Hide, 1953, 1958; Fowlis and Hide, 1965; Hide and Mason, 1975), and analyzed in details in Subsection 3.1. The setup, also used in (Gyüre et al., 2007) (see Subsection 3.3), consists of three concentric cylinders of radii $R_1 = 4.5$, $R_2 = 15.0$, and $R_3 = 20.3$ cm which is fixed on a rotating platform. The central container is filled up with a mixture of ice and water, the outermost one is regulated by an immersion heater, the working fluid in the middle annular region is tap water in the presented experiments. The main control parameters are the angular velocity $\Omega \in [1.54, 2.31]$ rad/s and imposed radial temperature difference $\Delta T \in [15.0, 40.5]$ °C in the dish. The height of the working fluid H is varied in a narrow range of 3.3-4.0 cm in order to warrant that the dynamics is well inside the geostrophic

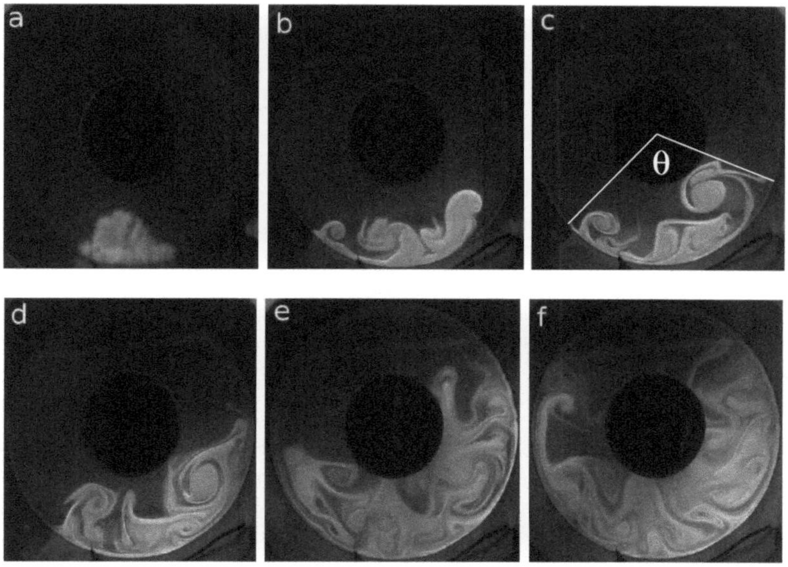

Figure 7. The spread of fluorescent dye as a function of time at parameters $\Delta T = 31.5$ °C, $\Omega = 1.62$ rad/s, $H = 3.4$ cm ($\mathrm{Ro}_T = 0.0209$). (a) 1 s, (b) 8 s, (c) 19 s, (d) 46 s, (e) 120 s, and (f) 154 s after the injection. The angle θ characterizes the zonal range covered by the cloud.

turbulent regime (Hide and Mason, 1975; Salmon, 1998). The convenient nondimensional combination Eq. (3) introduced in Section 2 as thermal Rossby number covers the range $\mathrm{Ro}_T \in [0.01, 0.09]$ with $\alpha \approx 2 \times 10^{-4}$ °C^{-1} for water, and $L = R_2 - R_1 = 0.105$ m.

Fluorescent dye (Sodium fluorescein, $C_{20}H_{10}O_5Na_2$) is used as passive tracer, and dispersion is evaluated by digital image processing (Jánosi et al., 2010). Figure 7 illustrates a typical experiment, where the zonal range of the tracer cloud can be characterized by the time dependent angle $\theta(t)$. Furthermore, the total number of pixels $n(t)$ above a contrast threshold is also determined for each frame. In order to decrease the effects of inhomogeneities of UV illumination, an averaging over one revolution is performed prior to further processing. Note that while $\theta(t)$ is a simple linear measure of tracer dispersal, the time evolution of the total area $n(t)$ is affected by at least three processes with different weights at a particular set of parameters. Chaotic advection by the background flow field continuously increases the

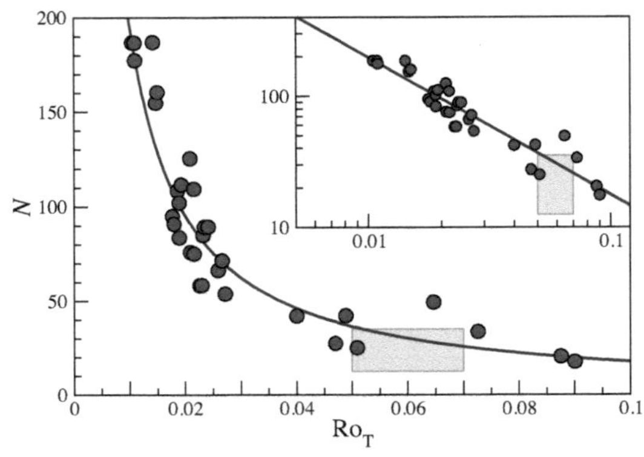

Figure 8. Number of revolutions N until the zonal spread of the dye cloud covers a whole circle as a function of thermal Rossby number Ro_T. The inset shows the same data on a double logarithmic scale. The continuous line is an inverse relationship $N \sim \mathrm{Ro}_T^{-1}$. Shaded rectangle indicates plausible range for the rotating Earth (Jánosi et al., 2010).

colored area, meanwhile local shearing contributes to mixing mostly between vortex walls and cores, however diffusional thinning continuously decreases the fluorescent intensity. Therefore we do not expect that the zonal extent θ obeys a trivial relationship with a composite measure of total cloud size n. The experimental results, served as benchmark for subsequent numerical simulations in Jánosi et al. (2010), can be summarized as follows.

The dominating patterns clearly indicate strong irregular cyclonic and anticyclonic activity (see Fig. 7). It is widely accepted that this dynamics driven by the baroclinic instability reflects the most essential features of mid-latitude atmospheric flow (see Subsection 3.1). The general time evolution for both quantities $\theta(t)$ and $n(t)$ is found to be approximately linear: $\theta(t) = m_\theta(\mathrm{Ro}_T)t + \theta_0$ and $n(t) = m_n(\mathrm{Ro}_T)t + n_0$ with slopes m_θ and m_n depending on the control parameter Ro_T. Linear zonal growth can be followed up to $\theta = 360°$. The growth of total colored area proceeds also to a given time, where a crossover to a constant apparent cloud size is visible (Jánosi et al., 2010). As for the dependence on the control parameter, we obtained a linear relationship for slopes m_θ ($m_\theta \sim \mathrm{Ro}_T$), and a power law $m_n \sim \mathrm{Ro}_T^\delta$ with $\delta \approx 2$ (Jánosi et al., 2010) . Note that we could not find any theoretical prediction for such behavior in the literature available for us.

An interesting quantity to be compared with real observations is the encompassing time t^* expressed as a dimensionless number $N = t^*\Omega/2\pi$, which is the number of revolutions until the leading and tailing edges of the tracer cloud cross the same "longitude" in the rotating annulus. In order to minimize the effect of different initial configurations, the measured temporal interval of $t(\theta_0 \to 360°)$ is extrapolated to $t(0 \to 360°)$ assuming the same average zonal spreading velocity (here θ_0 is the initial zonal extent of the dye cloud after the injection). The result is plotted in Fig. 8. The relationship is clearly an inverse function $N \sim \mathrm{Ro}_T^{-1}$: the stronger the convective drive the lower the number of revolution until a "hemispheric" encompassing.

If we intend to compare experimental N values with observations, we need an estimate for the thermal Rossby number Ro_T for the rotating Earth. This is not trivial, because of the differential rotation of the Earth (the Coriolis parameter changes with the latitude) and the different thermal boundary conditions (distributed heating of the atmosphere). Still, when we assume that the laboratory setup adequately models mid-latitude atmosphere between 30°-70°, an approximate mean temperature difference can be $\Delta T \approx 25 - 30°\mathrm{C}$. Relevant length scales are $H = 10$ km, $L = 4500$ km, a mean Coriolis parameter [replacing 2Ω in Eq. (3)] is $f \approx 10^{-4}\ \mathrm{s}^{-1}$, and an average coefficient of volumetric expansion for air is $\alpha \approx 4 \times 10^{-3}\ °\mathrm{C}^{-1}$. An estimated thermal Rossby number [see Eq. (3)] should fall in the range [0.05-0.07]. The range of plausible values for N can be estimated from numerical simulations of passive tracers in reanalysis wind fields (Jánosi et al., 2010). Obtained results represented by the shaded rectangle in Fig. 8 are consistent with the laboratory experiments and thus with the underlying assumptions.

3.3 Asymmetric temperature fluctuations in the atmosphere

Although it is clear that atmospheric dynamics is inherently nonlinear and irreversible, several studies have been suggested that the synoptic-scale activities may be approximated by linear dynamics. Accordingly, strong correlations characterize the dynamics for short time intervals that permits a satisfactory description of atmospheric parameter changes as a low order linear autoregressive process (Király and Jánosi, 2002). Such processes are symmetric under time reversal leading to the 'reversibility paradox' (Loschmidt, 1876): how to deduce an irreversible process from time-symmetric dynamics? Both macroscopic irreversibility and microscopic time-reversal symmetry are well-accepted principles in physics, with solid observational and theoretical support, yet they seem to be in conflict; hence the paradox.

The lack of time-reversal symmetry of temperature fluctuations in labo-

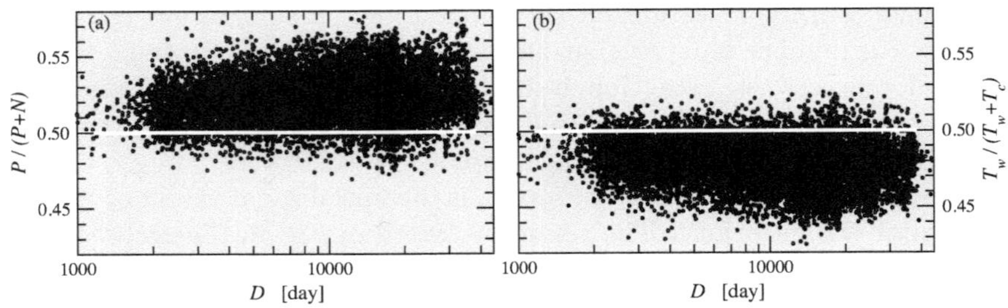

Figure 9. Two asymmetry measures as a function of record length D. (a) The number of positive daily mean temperature steps over the total number of steps $P/(P+N)$, and (b) the ratio $T_w/(T_w+T_c)$ where T_w and T_c denote the mean one-day temperature change of warming and cooling averaged over the total length of a given record. The results for 13380 terrestrial meteorological stations are plotted. (Bartos and Jánosi , 2005; Gyüre et al., 2007; Kiss and Jánosi, 2010)

ratory experiments on turbulent thermal convection is relatively well known (Belmonte and Libchaber, 1996), where the statistics of time derivative of temperature records is found to be an appropriate tool to characterize the dynamics of the flow. The temporal asymmetry is simply quantified by the skewness of magnitude probability density functions. The observed behavior is that the skewness has a positive value at the cold (top) boundary, it changes sign at around the border of the cold thermal boundary layer, and it is increasingly negative for larger distances (Belmonte and Libchaber, 1996). The question naturally arises: is there a similar behavior expected in the atmospheric boundary layer, where thermal convection is also a fundamental constituent of dynamics?

A comprehensive analysis of 13380 daily mean temperature time series from the Global Daily Climatology Network (GDCN, compiled by the National Climatic Data Center) resulted in an apparent time reversal asymmetry considering one day temperature differences. Figure 9a shows that the number of warming steps is definitely larger than the number of cooling steps at almost every terrestrial locations (mostly in the northern midlatitude band) irrespective of the length of the record. Here the simple asymmetry measure $P/(P+N)$ is used, where P and N denote the number of positive and negative steps. This strong asymmetry is necessarily compensated by larger average values for the cooling steps (Fig. 9b), here T_w and T_c are the mean warming and cooling step magnitudes at a given mea-

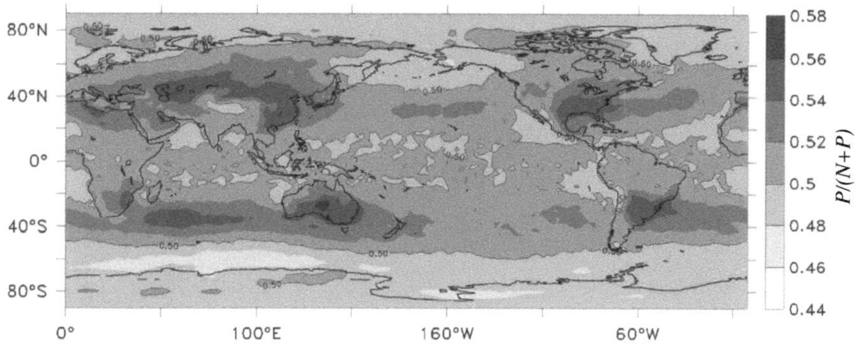

Figure 10. A map of the one-day asymmetry measure $P/(P+N)$, based on the surface daily mean temperature of the NCEP–NCAR reanalysis for the time period 1948–2006. (Ashkenazy et al., 2008; with permission)

suring site. The correlation between the two ratios (warming step number over cooling step number, and mean cooling step size over mean warming step size) is very strong, the computed correlation coefficient has a value of $r = 0.984$. The observed asymmetry is a consequence of higher order nonlinear correlations in the time series demonstrated by Kiss and Janosi (Bartos and Jánosi, 2006; Kiss and Jánosi, 2010).

In a comprehensive follow-up study, Ashkenazy et al. (2008) evaluated 59 years of the NCEP/NCAR (National Centers for Environmental Prediction, and National Center for Atmospheric Research) reanalysis data, including a global, multilevel coverage of temperature with a spatial resolution of 2.5×2.5. The measure of asymmetry they used is based on warming- and cooling-step numbers, too. Apart from the vicinity of the equator, daily surface-temperature fluctuations obey a significant time asymmetry on both hemispheres (Figure 10). The general behavior is that large cooling steps are followed by slower gradual warming. This is a statistical pattern, the temperature records rarely exhibit a very clear sawtooth pattern. The asymmetry changes sign at latitudes around ±60°. At higher latitudes, sudden warming jumps are followed by slower gradual cooling. The asymmetry fades away as a function of altitude indicating clearly the role of atmospheric boundary layer. The asymmetry also fades away when longer-time differences are evaluated, such short-time 'memory' (persistence) of weather phenomena is known to have a characteristic time scale of 4-5 days.

Based on their careful analysis and detailed comparisons with annual

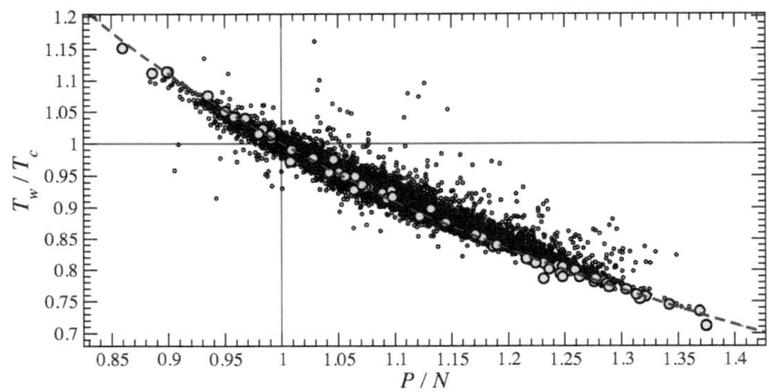

Figure 11. Correlation plot between step number ratio P/N and average step size ratio T_w/T_c for one-day temperature changes. Small cyrcles: 13380 terrestrial weather stations. Large gray circles: laboratory experiments. Dashed line indicates the perfect stationarity condition. (Gyüre et al., 2007)

mean transient heat flux and the frequency of low pressure events (storms), Ashkenazy et al. (2008) formulated a convincing explanation of the observed time reversal asymmetries in daily mean temperature changes. The rapid cooling seen in the mid-latitudes and the rapid warming seen in the high latitudes must be associated with the cold or warm fronts that propagate toward low/high latitudes, correspondingly. After the passage of a cold or warm front, the temperature relaxes to the equilibrium atmospheric temperature, where the (possibly small scale convective, diffusive and radiative) relaxation processes are relatively slow compared to the front propagation. At the mid-latitudes, the temperature slowly relaxes after the passage of the cold front to the warmer equilibrium temperature; at the high latitudes, the temperature slowly relaxes after the passage of the warm front to the colder equilibrium temperature.

Since the main message of this Chapter is that a thermally driven rotating annulus reproduces the basic physics of the mid-latitude atmosphere, it is quite plausible to perform an experimental check of temperature fluctuations. Gyüre et al. (Gyüre et al., 2007) implemented temperature measurements in the classic rotating annulus sketched in Figure 4b. Angular velocities in the range 1.88-4.71 s$^{?10}$ at two temperature gradients were imposed: $T_{w0} = 35.0 \pm 0.1$°C and 40.0 ± 0.1°C outside, melting ice ($T_{c0} \approx 4$°C) inside. Two corotating Ni-NiCr thermocouples fixed at the end of thin wires of diameter 0.5 mm were sampled at a rate of $\Delta t = 3.0$ s, which is approximately one sample per revolution, in analogy with the meteorological records. The height of the sensors was fixed at 3 mm from the bottom; the radial positions were changed in the experiments running 5?6 hrs each. In the non-rotating control experiments the presence of simple radial convection was checked, and the typical flow speeds obtained were between 0.5 and 1.0 mm/s. The imposed parameter range is deeply in the dynamical regime of irregular wave patterns (cf. Figure 5), similarly to the mid-latitude atmosphere.

The experimental temperature records were evaluated in the same way as the meteorological data. The same robust asymmetries in the statistics of temperature changes were observed; the results are shown in Figure 9. The apparent data collapse onto the curve of strict stationarity in Figure 11 (dashed line) is somewhat surprising for the meteorological data, which means that systematic baseline drifts (urbanization, changing land use, global warming, etc.) are hardly visible in the records. The simple experimental setup cannot model many fundamental aspects of the real atmosphere, such as the strong density stratification (compressibility), the distributed differential heating by insolation, or the latitude dependent strength of the Coriolis parameter. Still, the marked violation of time reversal symmetry in the experimental tank further strengthens the physical analogy between the geostrophic turbulence observed in the troposphere and the differentially heated rotating annulus.

3.4 Interdecadal climate variability in the laboratory

Atlantic Multidecadal Oscillation (AMO) – a term coined by Kerr (2000) – or as it is more accurately referred to, 'Atlantic Multidecadal Variability' (AMV) is a robust temperature signal in the interdecadal frequency band of the paleoclimate and instrumental records of the North Atlantic region, stretching back to centuries. In the past years, a consensus has been emerging within the community that at least two different, more or less independent processes contribute to this variability: one of 20-30-year and another of 50-70-year characteristic timescales (Dong and Sutton, 2005; Vellinga and Wu, 2004). More radically, it has been suggested (by the authors) that this interdecadal spectral band may not be special by any means after all, but can be interpreted as a realization of the same multi-scale long-range-correlated stochastic dynamics that yields proper fits to the anomalies of much smaller timescales as well (Vincze and Jánosi, 2011).

The importance of AMV's contribution to global climate is immediately visible when comparing the two time series of Figure 12 The top curve shows the instrumental record of the so-called 'AMO index' (AMOI), introduced by Enfield (Enfield et al., 2001), that is defined as the 121-month running

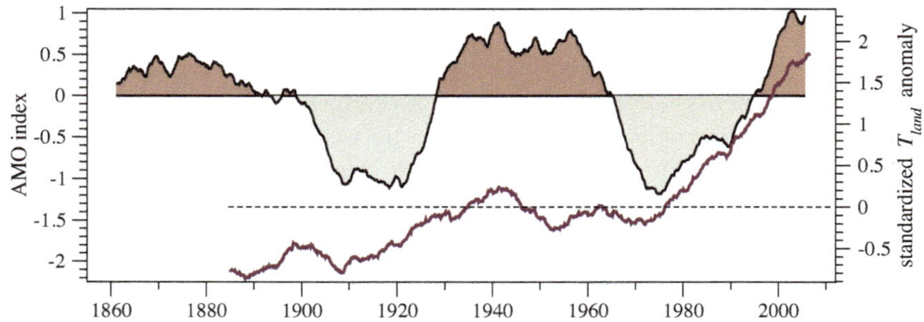

Figure 12. AMO index (left, thin black line) and standardized global mean temperature anomaly over land T_{land} (right, thick black line) shifted downwards for a better visualization. Both records are smoothed by a 121 months running mean filter. Sources: http://www.esrl.noaa.gov/, and http://www.ncdc.noaa.gov/.

mean of sea surface temperature (SST) anomalies, averaged over the North Atlantic basin. The 'warm' 1940s (positive AMOI), and the 'cold' 1970s (negative AMOI) – when climate scientists envisioned the signs of a new ice age – can be clearly distinguished. It is interesting to note that the standardized global mean temperature record over land T_{land} shown by the bottom curve of Figure 12 (shifted downwards for visualization), after the same type of 121-month smoothing, appears to exhibit a rather robust correlation with the AMOI, superposed onto the increasing trend: clearly, AMV is one of the strongest components of mid-latitude climate variability. The phase of AMV also correlates with a certain spatial pattern of the North Atlantic SST field, as reported by Kushnir (Kushnir, 1994). In the AMOI > 0 intervals, SST anomalies are generally positive in the whole basin, except for the coast of Newfoundland, where a localized negative temperature anomaly appears. In the cold phases the sign reverses: an overall cold anomaly develops with a warm domain around Newfoundland.

Although the 'ingredients' of AMV may be numerous and complex, a surprisingly simple conceptual dynamical minimal model has been proposed by te Raa and Dijkstra (2002), which can reproduce several basic features of the 20-30-year 'mode' of the variability. Their model consists of a rotating rectangular sector of a uniform depth ocean and meridional SST gradient. Frankcombe et al. (2009) showed that a multidecadal oscillatory mode can be excited in this arrangement, by adding temporally and spatially correlated red noise forcing to the SST field, representing ocean-atmosphere interactions. This model is "minimal", or irreducible in the sense that if any of the three key components (rotation, meridional temperature gradient and thermal noise) is removed, the oscillatory mode can no longer be excited.

After learning about this conceptual numerical model of appealing simplicity from Henk *Dijkstra* in a summer school on climate variability, we decided to construct a laboratory experiment in which its implications can be tested (Vincze et al., 2012). In the apparatus the North Atlantic basin was represented by a rectangular acrylic tank, divided into three sectors by two internal vertical walls, as depicted in Figure 13. The central domain of length $L = 68$ cm and width $D = 25$ cm was filled up to height $H = 10$ cm with tap water. One of the side sectors – separated by a copper internal wall from the central domain – was packed full of melting ice, enough to keep the temperature in this separated compartment at (0 ± 0.1)°C for up to 5 hours (the average duration of our experimental runs). On the opposing vertical sidewall, an electric heating element was mounted, capable of releasing a maximum flux of 0.3 W cm^{-2}. These two heat sources provided the analogue of the meridional temperature gradient in our setup. The differential heating initiates a "sideways convective" flow (similarly to the rotating annulus experiments of the previous sections): in the absence of critical Rayleigh number, *any* temperature difference ΔT between the sidewalls is sufficient to do so. The apparatus was placed on a turntable, rotating at period $P = (3.0 \pm 0.05)$ s. Nine digital thermometers, placed into the top 1 cm of the working fluid were arranged into an equidistant grid and logged the 'sea surface' temperature values in the tank at a temporal resolution of 1 s^{-1}.

The third key component (besides temperature gradient and rotation) in the minimal model of Frankcombe et al. (2009) is the perturbative effect of a spatially correlated and temporally red noise-like anomalous heat flux at the water surface (representing 'weather'). This 'ingredient' was modeled by a halogen lamp of large infrared emission, mounted 50 cm above the water surface. These heat flux perturbations were on the order of 0.5°C all over the water surface (whereas the values of the 'merdidional' ΔT varied between 0.25 and 1.75°C). The lamp was turned on and off according to a stochastic sequence, controlled by a computer, with a mean time interval $m = 200$ s between subsequent actions, and standard deviation $\sigma = 50$ s. Thus the timescale of thermal forcing was approximately 400 s, or 130 P. Our control experiments demonstrated that indeed all three components are needed to excite low frequency variability in the set-up. If any one of lamp forcing, rotation or sidewall heating/cooling was absent, the low frequency

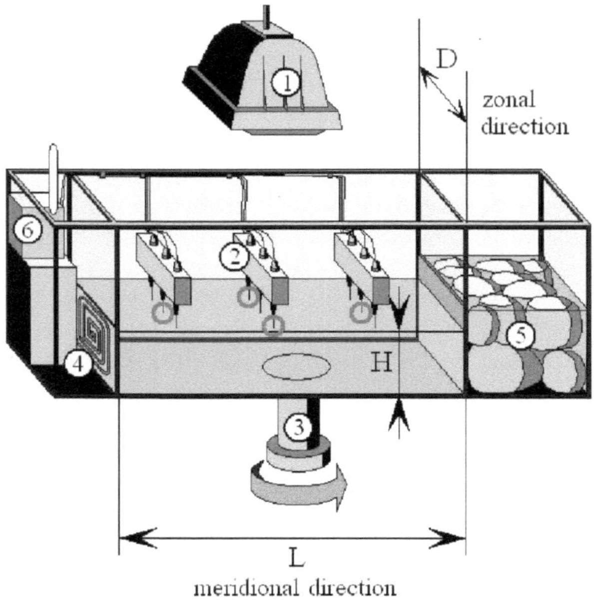

Figure 13. Schematic drawing of the setup. 1: lamp (for the surface heat flux perturbation), 2: digital temperature sensors (nine in total), 3: axis of rotation (the direction of rotation is also indicated), 4: electric heating module ("equator"), 5: the cooling sector packed full of ice ("polar region"), 6: radio transmitter for real time data acquisition. The geometric parameters L, D and H are indicated, together with the corresponding terminology ("zonal" and "meridional") (Vincze et al., 2012).

temperature oscillations disappeared (see Vincze et al. (2012) for details).

Next, using the average 'meridional' temperature difference $\langle \Delta T_M \rangle$, measured between the heated and cooled end of the tank as a control parameter, we evaluated the periods corresponding to the largest spectral amplitude for seven experimental runs at different values of side heating. The periods were acquired from the 'zonal' temperature difference anomalies (δT_Z).

In the minimal model, the period of the low frequency mode is set by the 'patch crossing time' i.e. the time it takes for a SST anomaly to drift across the North Atlantic basin in the zonal direction. In a thermally driven rotating system, an order-of-magnitude estimate of the drift velocity can

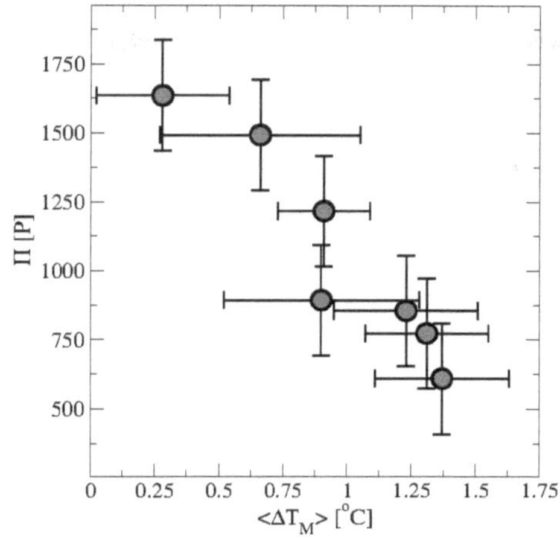

Figure 14. Period of the largest observed Fourier component of the band-pass filtered zonal temperature difference anomalies (δT_Z), as a function of the average meridional temperature difference (Vincze et al., 2012).

be obtained as written in equation (5) of Section 2. From here, one can conclude that the patch crossing time, and hence the period of the oscillatory mode scales with ΔT^{-1}. This assumption fairly agrees with our finding that a clearly decreasing trend can be observed, as visible in Figure 14.

As a "mechanistic indicator" associated with the *spatial pattern* of the multidecadal variability, Dijkstra et al. (2006) proposed to measure the phase lag between east-west and north-south temperature differences. Inspired by this idea we processed the "meridional" and "zonal" temperature difference anomaly signals (δT_M and δT_Z) accordingly. In order to quantify the phase shift between the two time series, cross correlation analysis was conducted with a maximal lag of 5000 s (1667 P); as a rule of thumb we chose the location of the first maximum at a positive lag as a measure of the phase shift.

As mentioned above, the mechanism of the 20-30 year mode of the multidecadal oscillation is supposedly based on the principle of quasi-geostrophic *thermal wind balance*, which enables a meridional temperature anomaly to initiate an anomalous zonal velocity perturbation. This provides an anomalous (perpendicular) component to the meridional overturning background flow, and by doing so, the temperature anomaly is getting pushed towards a meridional sidewall, yielding a zonal temperature gradient, and to the excitation of a meridional overturning anomaly (te Raa and Dijkstra, 2002).

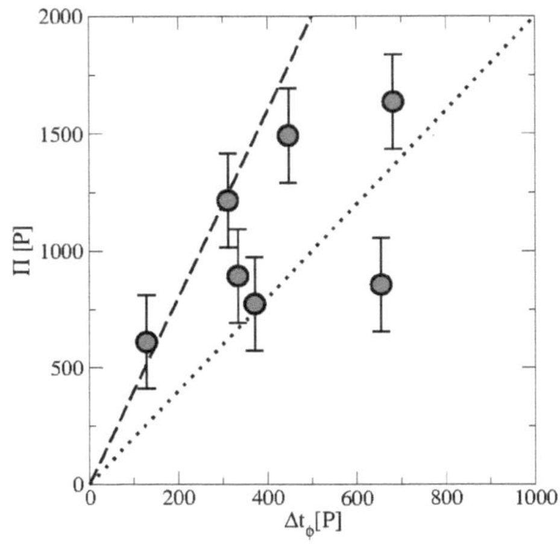

Figure 15. Period of the largest observed Fourier component of the band-pass filtered zonal temperature difference anomalies (δT_Z) as a function of the phase lag between the meridional and zonal temperature difference anomalies (obtained via cross correlation analysis). The dotted line represents $y = 2x$, the dashed line marks $y = 4x$. The fact that most of the data points lie between the two lines, implies positive correlation (Vincze et al., 2012).

This interplay between temperature and velocity anomalies drives the observed pattern formation in the system and sets the timescale of the oscillation. As a consequence of this reasoning, one would expect a correlation between the above discussed zonal-meridional phase lags and the period of the oscillation itself. Figure 15 shows this dependence for seven experimental runs. For each run, the phase lag value was obtained via the aforementioned cross-correlation method. Although the results are not exactly conclusive, yet a clear trend is present: most of the observed phase lags are found to be between quarter and half a period (see dotted and dashed lines in Figure 15). Naturally, for a clear oscillatory mode with one single frequency and one persistent, permanently cycling surface patch (a "thermal Rossby wave"), the phase lag between zonal and meridional temperature differences would be exactly one fourth of a period. However, for a "lamp noise"-induced dynamics present here, it is not surprising at all that the spatial and temporal behavior apperars to be more complex.

4 Concluding remarks

With this chapter, we intended to give the reader an overview on the role that laboratory experimentation has been playing in the development of meteorology, oceanography and climatology since the 18th century, and an insight into some of the current research topics. After discussing the past and the present, the question naturally arises of what kind of impact and contribution this research method can possibly have to the environmental sciences of the near future, marked by the 'digital revolution', the incredible growth of computing capacity available for numerical modeling.

One could of course argue in the spirit of Richard P. Feynman, saying that "experiment is the sole judge of scientific truth" (Feynman, 1967). However, in reference to atmospheric or oceanic dynamics we have to keep in mind that it is not possible to conduct actual experiments on planetary scales: in this sense laboratory experiments are merely incomplete models, just as much (or from many aspects, even more) as numerical simulations.

Thus the answer to the above question is twofold. Firstly, even compared to the best computers, nature itself still possesses the greatest possible computing power. In a properly set experiment, nonlinear phenomena which would require computationally intensive numerical codes to approximately implement, can be readily accessed in real time with practically infinite degrees of freedom. Continuing Feynman's above train of thought: "It always bothers me that according to the laws as we understand them today, it takes a computing machine an infinite number of logical operations to figure out what goes on in no matter how tiny a region of space and no matter how tiny a region of time". Indeed, treating such down-scaled models of continuous media as analog computers, or with the words of Carl-Gustaf *Rossby* (1898–1957) 'mechanical integrators' (Rossby, 1926) is one of the greatest strengths of experimental modeling. This is especially true for systems where wide ranges of spatial and temporal scales are heavily interconnected, and therefore a cut-off in resolution (inevitable in simulations) could significantly alter the observed dynamical behavior.

Secondly, experimental minimal models provide a superb test-bed to tune and validate large and complex 'Nimitz-class' general circulation models (GCMs) of the ocean and the atmosphere, designed for weather forecasts and global climate predictions. Systematic tests of these operational codes are especially hard to perform; separating the inaccuracies that arise due to faulty numerical implementation from the ones originating from our incomplete understanding of the processes of Earth's weather system (e.g. the details of cloud formation, the role of aerosols, etc.) poses a real challenge to researchers. In the laboratory-based systems, however, all governing equations and boundary conditions are fairly well known and can be adjusted or controlled precisely. Therefore all the errors and deviations that come up when simulating the flow in the experimental set-up using these GCMs cannot be blamed on the complexity of climate, but must be attributed to improper implementation.

As an example, we refer to the German Science Foundation's (DFG) *MetStröm* project (2008–2014), whose primary goal was to realize this very concept. A water-filled differentially heated rotating annulus served as a reference experiment and the model flow was simulated by five different working groups and GCMs, using various numerical approaches, solvers and subgrid-scale parametrization methods (Vincze et al., 2015). The results have uncovered some systematic discrepancies between the majority of the investigated numerical models and the experiment, and led to the refinement of their methods, hopefully contributing to better weather forecasts in the future.

Bibliography

Ashkenazy, Y., Y. Feliks, H. Gildor, and E. Tziperman, *Asymmetry of daily temperature records*, J. Atmos. Sci. **65**, 3327-3336, 2008.

Aubert, O., M. Le Bars, P. Le Gal, and P. S. Marcus, *The universal aspect ratio of vortices in rotating stratified flows: experiments and observations*, J. Fluid Mech. **706**, 34-45, 2012.

Bartos, I., and I. M. Jánosi, *Atmospheric response function over land: Strong asymmetries in daily temperature fluctuations*, Geophys. Res. Lett. **32**, L23820, 2005.

Bartos, I., and I. M. Jánosi, *Nonlinear correlations of daily temperature records over land*, Nonlin. Proc. Geophys. **13**, 571-576, 2006.

Belmonte, A., and A. Libchaber, *Thermal signature of plumes in turbulent convection: the skewness of the derivative*, Phys. Rev. E, **53**, 4893-4898, 1996.

Borchert, S., U. Achatz, and M. D. Fruman, *Gravity wave emission in an atmosphere-like configuration of the differentially heated rotating annulus experiment*, J. Fluid Mech. **758**, 287-311, 2014.

Boschan, J., M. Vincze, I. M. Jánosi, and T. Tél, *Nonlinear resonance in barotropic-baroclinic transfer generated by bottom sills*, Phys. Fluids, **24**, 046601, 2012.

Chen, C. F., D. G. Briggs, and R. A. Wirtz, *Stability of thermal convection in a salinity gradient due to lateral heating*, Int. J. Heat Mass Transport, **14**, 57-65, 1971.

Coman, M. A., Griffiths, R. W., and Hughes, G. O., *Sandström's experiments revisited*, J. Marine Res., **64**, 783-796, 2006.

Delworth, T. L., and R. Greatbatch, *Multidecadal thermohaline circulation variability driven by atmospheric surface flux forcing*, J. Climate, **13**, 1481-1495, 2000.

Dijkstra, H. A., *On the interaction of SST modes in the North Atlantic Ocean*, J. Phys. Oceanogr., **36**, 286-299, 2006.

Dijkstra, H. A., te Raa, L. A., Schmeits, M. and Gerrits, J., *On the physics of the Atlantic Multidecadal Oscillation*, Ocean Dyn., **56**, 36-50, 2006.

Dong, B. and R. T. Sutton, *Mechanism of interdecadal thermohaline circulation variability in a coupled ocean-atmosphere GCM*, J. Climate, **18**, 1117-1135, 2005.

Ekman, V. W., *On dead water*, Norw. N. Polar Exped. 1893-1896: Sci Results, XV, Christiana, 1904.

Enfield, D. B., Mestas-Nunez, A. M. and Trimble, P. J., *The Atlantic Multidecadal Oscillation and its relation to rainfall and river flows in the continental U.S.*, Geophys. Res. Lett., **28**, 2077-2080, 2001.

Ewart, T. E., and W. P. Bendiner, *An observation of the horizontal and vertical diffusion of a passive tracer in the deep ocean*, J. Geophys. Res.: Oceans **86**, 10974-10982, 1981.

Feynman, R., *The Character of Physical Law*, MIT press, 1967.

Fowlis, W. W., and R. Hide, *Thermal convection in a rotating annulus of liquid: effect of viscosity on the transition between axisymmetric and non-axisymmetric flow regimes*, J. Atmos. Sci. **22**, 541-558, 1965.

Frankcombe, L. M., Dijkstra, H. A. and von der Heydt, A., *Sub-surface signatures of the Atlantic Multidecadal Oscillation*, Geophys. Res. Lett., **35**, L19602, 2008.

Frankcombe, L. M., Dijkstra, H. A., and von der Heydt, A., *Noise-induced multidecadal variability in the North Atlantic: excitation of normal modes*, J. Phys. Oceanogr., **39**, 220-233, 2009.

Frankcombe, L. M., von der Heydt, A., and Dijkstra, H. A., *North Atlantic Multidecadal Variability: An investigation of dominant time scales and processes*, J. Climate, **23**, 3626-3638, 2010.

Fultz, D., *A preliminary report on experiments with thermally produced lateral mixing in a rotating hemispherical shell of liquid*, J. Atmos. Sci. **6**, 17-33, 1949.

Fultz, D., *On the possibility of experimental models of the polar-front wave*, J. Atmos. Sci. **9**, 379-384, 1952.

Fultz, D., R. R. Long, G. V. Owens, W. Bohan, R. Kaylor, and J. Weil, *Studies of thermal convection in a rotating cylinder with implications for atmospheric motions*, Meteor. Monogr. Amer. Meteor. Soc. **4**, 1-104, 1959.

Ghil, M., P. L. Read, and L. A. Smith, *Geophysical flows as dynamical systems: The influence of Hide's experiments*, Astron. Geophys. **51**, 28-35, 2010.

Gouillart, E., O. Dauchot, J.-L. Thiffeault, and S. Roux, *Open-flow mixing: experimental evidence for strange eigenmodes*, Phys. Fluids, **21**, 023603, 2009.

Gyüre, B., I. Bartos, and I. M. Jánosi, *Nonlinear statistics of daily temperature fluctuations reproduced in a laboratory experiment*, Phys. Rev. E, **76**, 037301, 2007.

Hide, R., *Some experiments on thermal convection in a rotating liquid*, Q. J. Roy. Met. Soc. **79**, 161, 1953.

Hide, R., *An experimental study of thermal convection in a rotating liquid*, Phil. Trans. Roy. Soc. Lond. A, **250**, 441-478, 1958.

Hide, R., and P. J. Mason, *Sloping convection in a rotating fluid*, Adv. Phys. **24**, 47-100, 1975.

Holton, J. R., *An introduction to dynamic meteorology*, 4th ed., Academic Press, Burlington, 2004.

Jánosi, I. M., P. Kiss, V. Homonnai, M. Pattantyús-Ábrahám, B. Gyüre, and T. Tél, *Dynamics of passive tracers in the atmosphere: laboratory experiments and numerical tests with reanalysis wind fields*, Phys. Rev. E, **82**, 046308, 2010.

Kaplan, A., Cane, M., Kushnir, Y., Clement, A., Blumenthal, M. and Rajagopalan, B., *Analyses of global sea surface temperature 1856-1991*, J. Geophys. Res. **103**, 567,589, 1998.

Kerr, R. A., *A North Atlantic climate pacemaker for the centuries*, Science, **288**, 1984-1985, 2000.

Király, A., and I. M. Jánosi, *Stochastic modeling of daily temperature fluctuations*, Phys. Rev. E, **65**, 051102, 2002.

Kiss., P., and I. M. Jánosi, *Time-asymmetric fluctuations in the atmosphere: daily mean temperatures and total-column ozone*, Phil. Trans. R. Soc. A, **368**, 5721-5735, 2010.

Klein, M., T. Seelig, M. V. Kurgansky, A. Ghasemi V., I. D. Borcia, A. Will, E. Schaller, C. Egbers, and U. Harlander, *Inertial wave excitation and focusing in a liquid bounded by a frustum and a cylinder*, J. Fluid Mech. **751**, 255-297, 2014.

Kundu, P. K., and Cohen, I. M., *Fluid Mechanics*, 4th revised ed., Academic Press, 2008.

Kushnir, Y., *Interdecadal variations in North-Atlantic sea-surface temperature and associated atmospheric conditions*, J. Climate, **7**, 141-157, 1994.

Lorenz, E. N., *The mechanics of vacillation*, J. Atmos. Sci. **20**, 448-465, 1963.

Lorenz, E. N., *The Nature and Theory of the General Circulation of the Atmosphere*, World Meteorological Organization, 1967.

Loschmidt, J., *Über den Zustand des Wärmegleichgewichtes eines Systems von Körperen mit Rücksicht auf die Schwerkraft, 1. Teil*, Sitzungsber. Kais. Akad. Wiss. Wien, Math. Naturwiss. **73**, 128-142, 1876.

Mann, E., Z. Zhang, S. Rutherford, R. S. Bradley, M. K. Hughes, D. Shindell, C. Ammann, G. Faluvegi, and F. Ni, *Global signatures and dynamical origins of the little ice age and medieval climate anomaly*, Science, **326**, 1256-1260, 2009.

Marsigli, C. L. F., *Histoire physique de la mer*, translated by Leclerc, Amsterdam, 1725.

Mitchell, D. M., L. Montabone, S. Thomson, and P. L. Read, *Polar vortices on Earth and Mars: A comparative study of the climatology and variability from reanalyses*, Q. J. Roy. Met. Soc. **141**, 550-562, 2015.

Oster, G., and M. Yamamoto, *Density gradient techniques*, Chem. Rev., 1963, 63 (3), pp 257-268, 1963.

Ou, H., *A minimal model of the Atlantic Multidecadal Variability: its genesis and predictability*, Climate Dynamics, **38**, 775-794, 2012.

Raa, te L. A., and H. A. Dijkstra, *Instability of the thermohaline circulation on interdecadal time scales*, J. Phys. Oceanography, **32**, 138-160, 2002.

Randriamampianina, A., *Inertia gravity waves characteristics within a baroclinic cavity*, Compt. Rend. Mecanique, **341**, 547-552, 2013.

Rossby, C. G., *On the solution of problems of atmospheric motion by means of model experiments*, Mon. Weather Rev., **54**, 237-240, 1926.

Salmon, R., *Lectures on Geophysical Fluid Dynamics*, Oxford University Press, Oxford, 1998.

Sandström, J. W., *Dynamische Versuche mit Meerwasser (Dynamical experiments with seawater)*, Annal. Hydrogr. Marit. Meteorol. **36**, 6-23, 1908.

Sinha, B. and Topliss, B., *A description of interdecadal time-scale propagating North Atlantic sea surface temperature anomalies and their effect on winter European climate, 1948-2002*, J. Climate, **19**, 1067-1079, 2006.

Sommerer, J. C., H. C. Ku, and H. E. Gilreath, *Experimental evidence for chaotic scattering in a fluid wake*, Phys. Rev. Lett. **77**, 5055-5058, 1996.

Stommel, H., *Thermohaline convection with two stable regimes of flow*, Tellus, **13**, 224-230, 1961.

Sutton, R. T., and Hodson D. L. R., *Atlantic Ocean forcing of north American and European summer climate*, Science, **309**, 115-118, 2005.

Taylor, G. I., *Stability of a viscous liquid contained between two rotating cylinders*, Phil. Trans. Roy. Soc. A, **223**, 289-343, 1923.

Vallis, G. K., *Atmospheric and Oceanic Fluid Dynamics - Fundamentals and Large-Scale Circulation*, Cambridge University Press, Cambridge, 2006.

Vellinga, M. and P. Wu, *Low-latitude freshwater influence on centennial variability of the Atlantic thermohaline circulation*, J. Climate, **17**, 4498-4511, 2004.

Vettin, F., *Experimentelle Darstellung von Luftbewegungen unter dem Einfluss von Temperatur-Unterschieden und Rotations-Impulsen*, Z. Meteor. **1**, 227-230, 1884.

Vincze, M. and I. M. Jánosi, *Is the Atlantic Multidecadal Oscillation (AMO) a statistical phantom?* Nonlin. Processes Geophys., **18**, 469-475, 2011.

Vincze, M., A. Várai, E. Barsy, and I. M. Jnosi, *The effect of a localized geothermal heat source on deep water formation*, Nonlin. Processes Geophys. **18**, 841-847, 2011.

Vincze, M., I. M. Jánosi, E. Barsy, T. Tél, and A. Várai, *An experimental study of the Atlantic variability on interdecadal timescales*, Nonlin. Processes Geophys. **19**, 335-343, 2012.

Vincze, M., S. Borchert, U. Achatz, Th. von Larcher, M. Baumann, C. Liersch, S. Remmler, T. Beck, K. Alexandrov, Ch. Egbers, J. Froehlich, V. Heuveline, S. Hickel, and U. Harlander, *Benchmarking in a rotating annulus: a comparative experimental and numerical study of baroclinic wave dynamics*, Meteorol. Z. **23**, 611-635, 2015.

Walker, J. M., *Farthest north, dead water and the Ekman spiral*, Weather, **46**, 158-164, 1991.

Hepeng Zhang, H., B. King, and H. L. Swinney, *Resonant generation of internal waves on a model continental slope*, Phys. Rev. Lett. **100**, 244504, 2008.

Individual Particle Based Description of Atmospheric Dispersion: a Dynamical Systems Approach

Tímea Haszpra* and Tamás Tél*,†

* MTA–ELTE Theoretical Physics Research Group, Pázmány P. s. 1/A, H-1117 Budapest, Hungary

† Department of Theoretical Physics, Eötvös Loránd University, Pázmány P. s. 1/A, H-1117 Budapest, Hungary

Abstract We argue that a proper treatment of material dispersion should be based on individual particle tracking using realistic size and density. The effect of turbulent diffusion and the scavenging of particles by precipitation are shown to be treatable as stochastic perturbations of the deterministic Newtonian equation of motion. This approach enables one to investigate the chaotic aspects of particle dispersion by means of dynamical systems concepts. Topological entropy is shown to be in this context the growth rate of material lines, which can be considered to provide a novel characterization of the state of the atmosphere. The deposition process is found to be well characterizable by the escape rate (being a measure of the strength of the exponential decay of the number of particles not yet reached the surface), which might depend on local turbulence and rain intensity. The variability of the dispersion process due to the difference between different meteorological forecasts within an ensemble forecast are also illustrated. Examples are taken from volcanic eruptions and the Fukushima accident.

1 Introduction

The concepts of chaos theory apply to any nonlinear system. Nowadays they are also widely used in different treatments of climate dynamics, as some chapters of this book also illustrate. The most appropriate appearance of dynamical systems theory is in conceptual climate models since chaos is basically a low-dimensional phenomenon. Primarily, it is a feature of temporal dependence without any spatial extension. Chaos is thus a property of systems describable by ordinary differential equations. The often heard statement that weather is chaotic should therefore be interpreted

A. Provenzale et al. (Eds.), *The Fluid Dynamics of Climate*,
CISM International Centre for Mechanical Sciences
DOI 10.1007/ 978-3-7091-1893-1_4

in a symbolic sense: weather, in which spatial features are essential, is more complex than chaotic. It is basically turbulent in accordance with the fact that weather is described by the partial differential equations of hydrodynamics. (In spite of the differences between chaos and turbulence, some features might be in common, like e.g. the fact that both phenomena are unpredictable.) As a consequence, all aspects of climate dynamics related to essential spatial features and requiring thus a description in terms of partial differential equations are more complicated than chaotic.

There is one class of phenomena, relevant both in weather- and climate-related contexts, namely the dispersion of particles, which is a chaotic process. This is so because the traditional description of any flow occurs in the Eulerian picture, and implies the determination of the velocity field. The advection of particles can, however, be treated as a phenomenon sitting on top of the Eulerian description. Particles basically follow the fluid velocity at their instantaneous position, and their motion, a Lagrangian feature, can be considered as one *driven* by the Eulerian flow. Particle trajectories, which are functions of time only, can thus basically be obtained as solutions of ordinary differential equations containing the velocity field (which itself fulfills a partial differential equation) as a known input function. In this sense advection is a clearly chaotic phenomenon as the term "chaotic advection" (Aref, 1984) expresses.

In the last decades the demand for precise tracking and forecasting of atmospheric pollutants has increased due to the growing interest in environmental problems and, consequently, to the requirements for detailed prediction of health and economic hazards. Recent events like volcanic eruptions (e.g. Mount St. Helens (1980), Pinatubo (1991), Eyjafjallajökull (2010)) and pollutant spreading from industrial accidents (e.g. Fukushima (2011)) underline the need for investigating pollutant dispersion in the atmosphere. Aerosol particles from different sources may be advected far away from their initial position and may cause air pollution episodes at distant locations. As discussed above, a Lagrangian description is needed in all cases when one is interested in pollutant trajectories[1].

A broad class of the currently used particle-tracking models tracks "ghost particles" (computational particles): any of these particles is assumed to be point-like, and carries an artificial mass which decays in time (e.g. HYSPLIT (Draxler and Hess, 2004), FLEXPART (Stohl et al., 2010)). The properties of these particles usually do not coincide with those of any real pollutant particle. In contrary, such particles typically follow the path of

[1] For describing pollutant concentrations, also Eulerian methods are available, in terms of advection-diffusion-sedimentation equations, but they are not based on individual particles, and thus cannot reflect chaos-related properties.

an air parcel (ideal passive tracer), gravitational settling does not have an effect on the motion of "ghost particles". The mass m (whose numerical value might be on the order of kilograms) attributed to the ghost particle is time-dependent and decreases due to dry and wet deposition according to the equation:

$$\frac{dm(t)}{dt} = -C(\mathbf{r}(t), t)m(t), \tag{1}$$

where $C(\mathbf{r}(t), t)$ is a location and time-dependent coefficient. It is the quantity C which is used to describe effects like gravitational settling, and dry or wet deposition.

In addition, in this approach a ghost particle is thought to be the center of mass of a large amount of adjacent pollutants. This assumption only holds if neighboring particles remain together forever. It is well-known, however, that advection is typically chaotic (Aref, 1984) and the nature of chaotic dynamics implies that an initially small, compact ball becomes rapidly deformed and strongly stretched. A measure of this strong deviation of neighboring particles is the so-called Lyapunov exponent (Ott, 1993). In a pictorial representation, the ball of real particles becomes deformed into a complicated, filamentary shape of large extent, see Figure 1. We should therefore conclude that the physical reality of "ghost particle" models is strongly questionable.

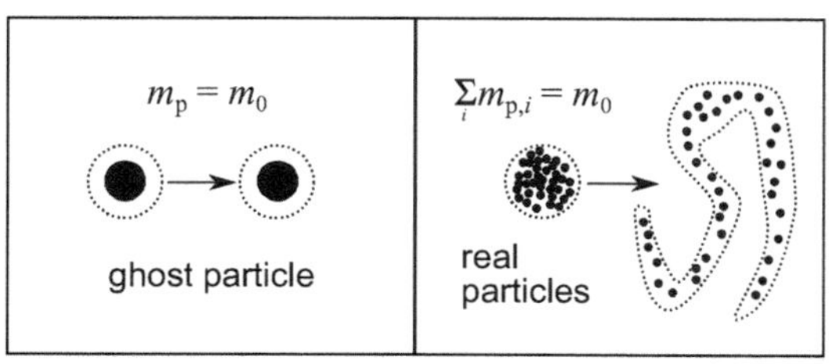

Figure 1. Comparison of the dispersion of a "ghost particle" and of real particles after some time, in the atmospheric context, after a few days. The total mass of the real particles of individual mass $m_{p,i}$ is equal to the mass $m_p = m_0$ of the ghost particle.

A faithful approach in the Lagrangian picture thus requires *real* particles: the particles in these models have fixed, realistic size and density (Heffter and Stunder, 1993; Searcy et al., 1998). However, the currently available models of this type do not take into account important effects like small-scale turbulence and the scavenging of particles by precipitation.

2 The RePLaT Lagrangian Dispersion Model

In order to cope with these effects, we expanded the validity of the available models (Heffter and Stunder, 1993; Searcy et al., 1998) and developed a relatively simple one, the Real Particle Lagrangian Tracking (RePLaT) model which is able to describe real physical processes reasonably well (Haszpra and Tél, 2013a). The model tracks individual aerosol particles with realistic size and density, and takes into account more processes than other "real particle" models. Since RePLaT describes the motion of realistic aerosol particles, it is also suitable for the investigation of the dispersion and deposition processes from a dynamical systems point of view.

The equation of motion for the particle trajectory $\mathbf{r}_{\mathrm{p}}(t)$ is derived from Newton's equation. The drag force depends on the particle Reynolds number that quantifies the relation of hydrodynamical and viscous acceleration due to the relative velocity between particle and fluid. Scale analysis reveals that the horizontal velocity of a small aerosol particle takes over the actual local wind speed practically instantaneously, whereas vertically the terminal velocity also has to be taken into account besides the vertical velocity component of air. Therefore a particle is advected by the wind components in the horizontal direction, and its vertical motion is, in addition, influenced by its terminal velocity w_{term} which depends on the size r and density ρ_{p} of the particle, as well as, on the density ρ and viscosity ν of the air at the location of the particle:

$$\frac{d\mathbf{r}_{\mathrm{p}}}{dt} = \mathbf{v} + w_{\mathrm{term}}\mathbf{n}, \tag{2}$$

where $\mathbf{n}$ is the vertical unit vector pointing upwards. For aerosol particles of size of at most 12 μm and with density about $\rho_{\mathrm{p}} = 2000$ kg/m^3 Stokes' law is valid during the full motion, and hence the terminal velocity is

$$w_{\mathrm{term}} = -\frac{2}{9}\frac{\rho_{\mathrm{p}} r^2}{\rho\nu} g. \tag{3}$$

Since the meteorological data utilized by the dispersion model have coarse resolution without resolving turbulent diffusion, the effect of small-scale turbulent diffusion on the particles is built into the model as a stochastic process. Within the planetary boundary layer the computation of the vertical turbulent diffusivity is based on the Monin–Obukhov similarity theory (see e.g. Dyer (1974); Troen and Mahrt (1986)), while the horizontal turbulent diffusivity is assumed to be constant in both the boundary layer and in the free atmosphere. The equation of motion completed by the impact of turbulent diffusion is

$$\frac{d\mathbf{r}_{\mathrm{p}}}{dt} = \mathbf{v} + w_{\mathrm{term}}\mathbf{n} + \xi\mathbf{D}, \tag{4}$$

where ξ is a random walk process and $\mathbf{D}$ represents the turbulent diffusion coefficient in the different directions, which might be location- and time-dependent.

A further novel feature of the model is the application of the impact of precipitation on individual particles by a random process which depends on precipitation intensity. For the parametrization of wet deposition, we have taken over what is used in the corresponding Eulerian approach (Seinfeld and Pandis, 1998). There the impact of wet deposition is taken into account via Eq. (1) with $C = k_\mathrm{w}$ called the wet deposition coefficient (or scavenging coefficient). This implies that after a short time Δt, locally a fraction $1 - \exp(-k_\mathrm{w}\Delta t) \approx k_\mathrm{w}\Delta t$ of mass becomes converted into wet material within the computational cell. We use this relationship to incorporate wet deposition into our model. We consider wet deposition as a random process that results in a particle being captured by a raindrop in time Δt with probability

$$p = 1 - \exp(-k_\mathrm{w}\Delta t). \tag{5}$$

Thereby the radius of the particle suddenly increases to the mean radius r_rain of raindrops which is on the order 100 µm or larger. The trajectory of the "new" particle (a particle that turned into a raindrop) is computed using the terminal velocity based on the new properties of the particle, and follows typically from the quadratic drag law. The "new" particle does not leave the atmosphere instantaneously, but falls through the air according to the equation of motion (4), with the terminal velocity

$$w_\mathrm{term} = -\frac{8}{3}\frac{\rho_\mathrm{rain} r_\mathrm{rain} g}{\rho C_\mathrm{D}} \tag{6}$$

of raindrops, where $C_\mathrm{D} = 0.4$ is the drag coefficient for spheres.

There are different parametrizations available for the typical radius of the raindrops (Sportisse, 2007). In RePLaT we use the Pruppacher–Klett parametrization (Pruppacher and Klett, 1997):

$$r_\mathrm{rain} = 0.488\, P^{0.21}, \tag{7}$$

where the unit of r_rain is mm and the unit of rain intensity P is $\mathrm{mm\,h^{-1}}$. Using this and other relations (Sportisse, 2007) the scavenging coefficient k_w appears in terms of rain intensity P as:

$$k_\mathrm{w} = 0.154\, P^{0.79}\ \mathrm{h}^{-1}. \tag{8}$$

For simplicity, the effect of wet deposition is taken into account only below the 850 hPa level.

These modifications due to turbulent diffusion and precipitation turn the equation of motion to a stochastic ordinary differential equation. The chaotic behavior found in this setting is thus obviously a kind of noisy chaos.

3 Data and Methods

As an input to the dispersion simulations, the reanalysis fields of the ERA-Interim database (Dee et al., 2011) and forecasts of the European Centre for Medium-Range Weather Forecasts (ECMWF) were used. The equations of motion of the particles are written in spherical coordinates in the horizontal, and in pressure coordinates in the vertical direction in agreement with the structure of the meteorological data used. The model solves the differential equations for the subsequent position of the particles by using the explicit Euler method. For dispersion taking place solely in the free atmosphere and for cases in which processes of the planetary boundary layer are also taken into account, the time step was chosen to be $\Delta t = 45$ min and 5.625 min, respectively. The meteorological data are available on a given latitude-longitude grid on different pressure levels in a given time resolution. The meteorological variables at the actual location of a particle are calculated using bicubic spline interpolation in the horizontal direction and linear interpolation in the vertical direction and also in time. For particle properties we take typical volcanic ash data (Johnson et al., 2012): r in the range of 1–10 µm and $\rho_\mathrm{p} = 2000$ kg/m^3.

Calculation of the pollutant concentration is based on the number of particles per cells of nearly equal horizontal area. This alternative grid is constructed in the following way: the meridional sides of a "rectangular" cell is equal for all latitudes, while the zonal sides of the cells vary with latitude so that the area of the cells is almost equal, see Figure 2 (for more details see Haszpra and Tél (2013b)).

4 Validation: the Fukushima Accident

In order to validate the RePLaT model we simulated the dispersion of two radioactive material, the aerosol-bound cesium-137 isotope (^{137}Cs) and the noble gas xenon-133 isotope (^{133}Xe) released during the accident of the Fukushima Nuclear Power Plant in the spring of 2011. Wind data are taken from the ERA-Interim database. The simulations took into consideration the processes of the boundary layer, like precipitation, and turbulent diffusion depending on the height, as described in Section 2. The data of local emission were taken from the literature (Stohl et al., 2012). The dispersion of the radioactive material was followed over several weeks. The

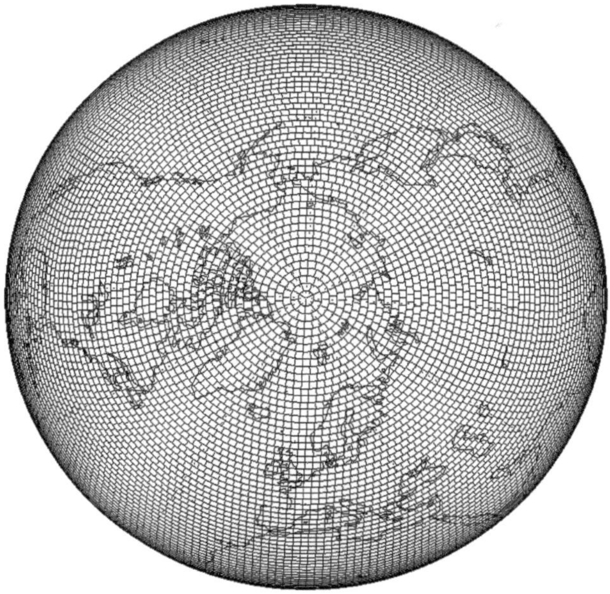

Figure 2. The nearly uniform-area grid used for the calculation of concentration. Here the side length of the cells is $\varepsilon \approx 1.5°$.

radius and density of ^{137}Cs carrier particles is estimated to be $r = 0.2$ µm and $\rho_\mathrm{p} = 1900\,\mathrm{kg/m^3}$ based on Stohl et al. (2012), while the noble gas component was treated formally as $r = 0$ particles (i.e. particles with zero terminal velocity) in Eq. (2). Figure 3 illustrates the geographical distribution of ^{137}Cs one week after the accident. The simulation shows that the particles are transported to the East over the Pacific Ocean. In mid March a fraction of the radioactive material is captured by the steering flow of two cyclones near Japan and at the coast of North America. These particles were lifted to the free atmosphere and reached even Europe about a week later.

Figure 4 shows the access radioactivity in the first few days after its arrival to different locations. It is remarkable that the arrival times of the pollution coincide reasonable well with the measured data (dashed lines). The deviations in the intensities might be related to the fact that the emission data are estimated a posteriori via retracking methods which are subject to considerable uncertainties (Stohl et al., 2012).

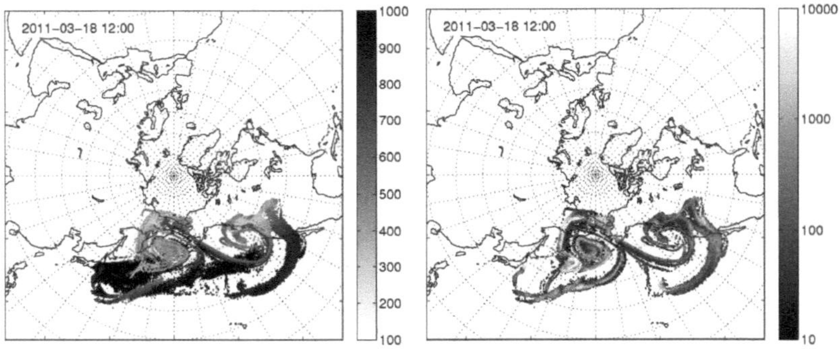

Figure 3. The dispersion of ^{137}Cs from the Fukushima accident on March 18, 2011. Initial conditions on March 11: particles initiated in a volume of $1° \times 1°$ area and height of about 300 m according to the *a posterior* estimation of Stohl et al. (2012). Left: altitude of the carrier $r = 0.2$ μm particles in hPa. Right: radioactivity concentration in air columns over the grid of Fig. 2 originating from ^{137}Cs in the unit of Bq/m^2.

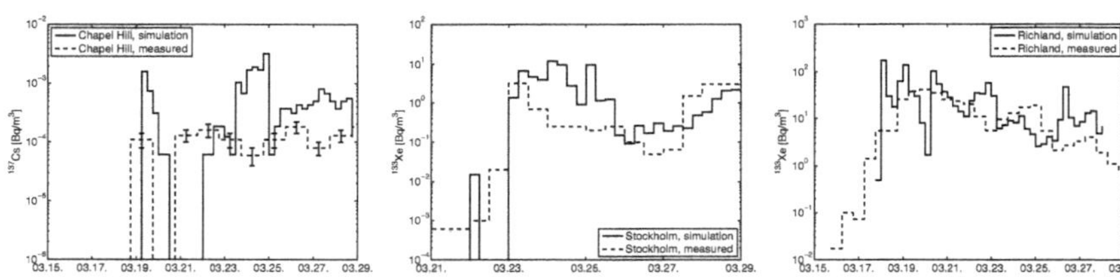

Figure 4. The access radioactive concentration of ^{137}Cs (left) and ^{133}Xe (middle and right) simulated by RePLaT (solid line) and the measured data (dashed line) as a function of the days of the year at Chapel Hill, Stockholm and Richland, respectively.

5 Topological Entropy

5.1 General Concepts

In dynamical systems theory, topological entropy is a measure of the complexity of the motion. In the most abstract setting, this quantity characterizes how the number of possible trajectories grows in time (Ott, 1993). The concept is most clearly accessible in periodically driven cases, where there exist unstable periodic orbits, so-called cycles, available for the dynamics. The temporal length of these cycles can be arbitrarily large. Moreover, the number N_t of all the unstable cycles of length t increases drastically, exponentially for large times t. The growth rate h defined by the relation $N_t \sim e^{ht}$ is called the topological entropy[2].

The existence of h is a basic property of chaos, so much that a possible definition of chaos is based on it: a system is chaotic if its topological entropy is positive (Ott, 1993; Tél and Gruiz, 2006). The unstable cycles form the skeleton of chaos, chaotic motion can be considered as random walk among the unstable cycles. The motion might temporarily approach one of the cycles. Since, however, the cycle in unstable, the trajectory can only remain in its neighborhood for a finite time and it approaches another one sooner or later. This is the origin of the irregular nature of chaotic dynamics.

A property of topological entropy which is easier to capture in measurements is that it also represents the growth rate of the length of line segments. A line segment of initial length L_0 is stretched more and more in the unstable direction of the dynamics. Let $L(t)$ denote the length of the line segment after time t. For two-dimensional systems it is proven (Newhouse and Pignataro, 1993) that after a sufficiently long time this length increases exponentially, and the growth rate is given by just the topological entropy, h, according to the relation

$$L(t) \sim e^{ht}, \tag{9}$$

valid for $t \gg 1/h$. The original definition based on unstable cycles and the one relying on the growth of line segments are equivalent in time-periodic dynamics. In aperiodic problems, however, only equation (9) can be used for the definition of topological entropy, and this is the approach we follow here in the context of atmospheric dispersion.

The topological entropy h is similar in spirit to, but different in value from, the (largest positive) Lyapunov exponent λ. A general inequality states (Ott, 1993; Tél and Gruiz, 2006) that

$$h \geq \lambda. \tag{10}$$

The difference lies in the fact that though the Lyapunov exponent is the rate of deviation between nearby trajectories, its definition reveals that the linear growth rate of the logarithm of the distance between a particle pair should be determined. In contrast, h is the rate of change of a length (and not of its logarithm). Inequality (10) is a consequence of the mathematical

[2]The terminology is motivated by Boltzmann's relation for the thermodynamical entropy S: $S = k_B \ln N$ known from statistical physics, where N is the number of states and k_B stands for Boltzmann's constant.

property that the logarithm of the average of a quantity is not the same as the average of the logarithm of the same quantity.

Technically, the evaluation of the topological entropy requires the monitoring of a large number of particles. This difficulty is, however, compensated by the fact that no smallness requirement or reshifting conditions are to be fulfilled (the latters are needed (Ott, 1993) for the Lyapunov calculation since the distance between the pair should always remain small). In particular, in flows represented on a grid, as in our case, the determination of the Lyapunov exponent faces the difficulty of being restricted to small scales. The determination of the topological entropy is based, however, on lengths exceeding by far the grid scale. The stretching filaments foliate regions with considerably different wind fields and are, therefore, natural candidates for providing a large scale characteristic of the atmosphere. Altogether, our experience shows that the numerical determination of the topological entropy appears to be straightforward and computationally rather cheap.

5.2 A Case Study

In the atmospheric context, the use of topological entropy is based on the general observation that any initially short material line becomes strongly stretched within a short time. We illustrate the usefulness of the concept with a case study within the RePLaT model. Wind data are taken from the ERA-Interim database again. For simplicity, first we consider ideal tracers ($r = 0$) within the free atmosphere where turbulent diffusion is negligible ($\mathbf{D} = 0$ in (4)), and no precipitation takes place ($P = 0$).

Fig. 5 illustrates the dispersion of an initially meridional line segment of $n = 2 \cdot 10^5$ particles with an initial length $L_0 = 3° \approx 333\,\text{km}$ (twice the resolution of the wind data) initialized between Great Britain and Scandinavia and followed for a period of 10 days. In the first days the particles are advected to Northeast towards the Scandinavian Peninsula, while the length of the filament increases. On the fourth day (Figure 5 left) the middle part of the filament is captured by a cyclone over Finland, while the easterly end also begins to spiral around another cyclone over Siberia. During the next few days (Figure 5 middle and right) the cyclones and anticyclones of the atmospheric flow fold, rumple and lengthen the filament more and more by stirring, and at the end of the observation period of 10 days the extent of the line segment becomes 6–7000 times longer than the initial length (Figure 6), and extends over Europe and quite a large area of Asia. It is also interesting, as can be read off from the grayscale of the right panel of Figure 5, that after 10 days the altitude of the particles spans to a range between about 300–1000 hPa, corresponding to about 9 km in altitude.

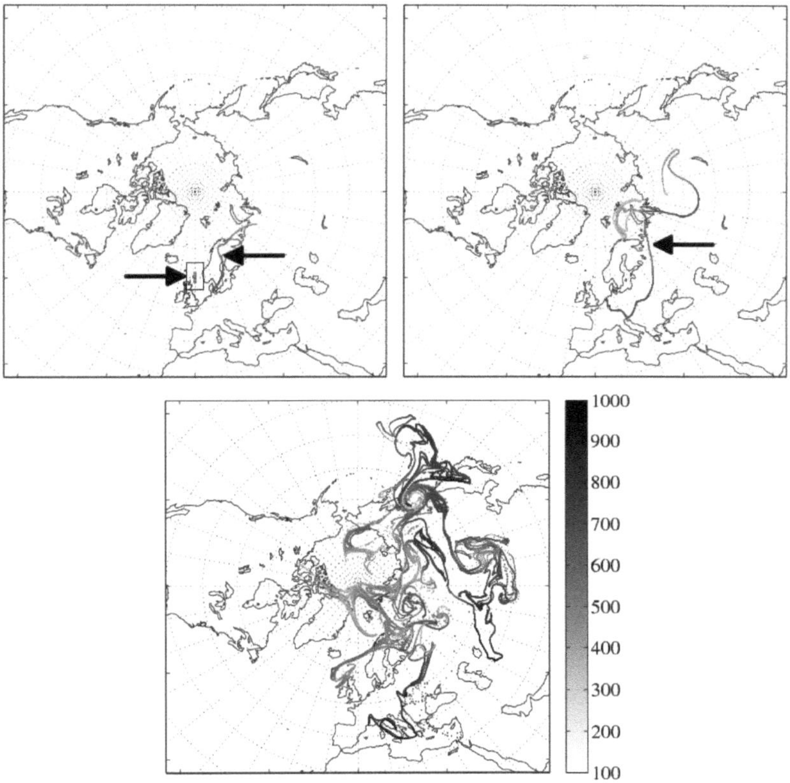

Figure 5. Stretching of a material line initialized at 00 UTC June 1, 2010. Initial conditions (surrounded by a box): $L_0 = 3° \approx 333$ km, $n_0 = 2 \times 10^5$ particles were distributed along a meridian on $p_0 = 500$ hPa, the center of the line was $\lambda_0 = 0°$, $\varphi = 60°$N. The panels show the location of the particles after 4, 6 and 10 days after the release, respectively. The arrows in the upper panels point toward the pollutant cloud. The altitude of the particles in pressure coordinates is marked by grayscale.

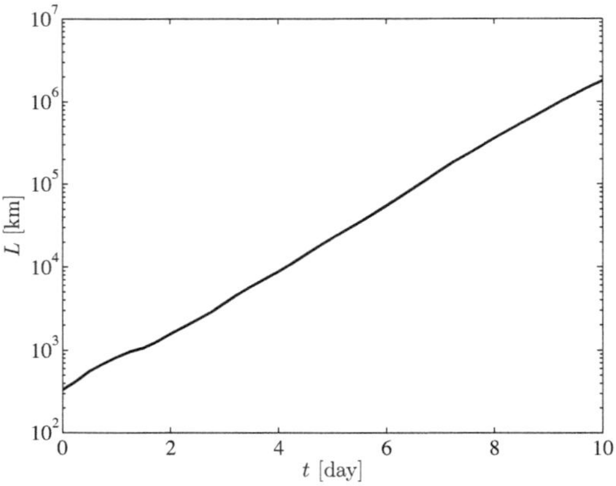

Figure 6. The length $L(t)$ of the material line shown in Figure 5 as a function of time over 10 days.

The length $L(t)$ of the filament at time t is computed as the sum of the horizontal distances between neighboring particle positions. It is clearly visible in Fig. 6 that the growth of $L(t)$ is exponential in time. The exponent, the topological entropy is found to be $h = 0.89$ day^{-1}. This implies that the total stretching factor after 10 days is $\exp(8.9) \approx 7330$, in harmony with the estimate above.

5.3 Geographical Distribution of Topological Entropy

One can observe quite remarkable differences in the topological entropy values depending on the initial geographical location and also on the particular season. To gain a systematic understanding, we extended our studies to small (up to $r = 5$ µm) aerosol particles and to different locations (Haszpra and Tél, 2011). We initialized material line segments oriented meridionally over the Globe, distributed on a grid: from 80°S to 80°N in 10° increments, and from 180°W to 180°E in 30° increments. The initial height is $p_0 = 500$ hPa. The topological entropy of each line segment is calculated from a 10-days tracking. The results for $r = 1$ µm particles obtained on the 1st of January 2010 are shown in Figure 7. The largest topological entropies (≈ 0.7 day^{-1}) appear in the midlatitudes, and can be attributed to the strong mixing and shearing effects of the cyclones. The smallest values (≈ 0.2 day^{-1}) can be found in the tropical belt.

The zonal average of these topological entropy values is of double-hill

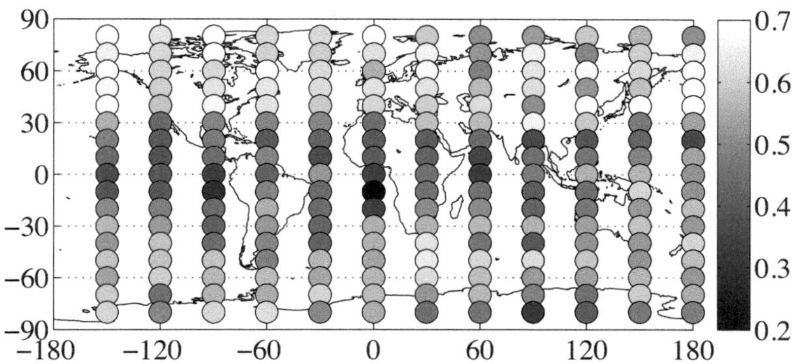

Figure 7. Geographical distribution of the 10-day-topological entropy of material lines of $r = 1$ µm particles. The material lines were distributed on the 500 hPa level on 01.01.2010 00 UTC in the center of the circles shown, their initial length was $L_0 \approx 333$ km.

shape (see Fig. 8), with somewhat larger values in the winter than in the summer hemisphere.

5.4 Remarks

We also carried out (Haszpra and Tél, 2013b) an investigation of the seasonal change of the topological entropy. To this end, at each geographical location, a line segment is initialized in every 10 days, then the temporal average of the topological entropy of three months is determined for December to February, March to May, June to August, and September to November. The largest values appear in the mid and high latitudes, mainly in the winter season of the hemisphere due to the strong mixing and shearing effects of cyclones. The zonally averaged topological entropy in the mid- and high latitudes (30°–80°) in the winter season of both hemispheres is somewhat larger than in the summer ($\overline{h}_{\rm w} - \overline{h}_{\rm s} \approx 0.06\,{\rm day}^{-1}$). This is in agreement with the fact that winters are more variable than summers because of the greater temperature gradient between the pole and the Equator (see also Figure 8). The difference between the winter and summer season is more significant on the Southern Hemisphere than on the Northern one (0.09 and 0.04 day^{-1}, respectively). The reason for this can be the difference in the proportion and location of oceans and continents. The average topological entropy was found to depend only slightly on the initial altitude of the particles.

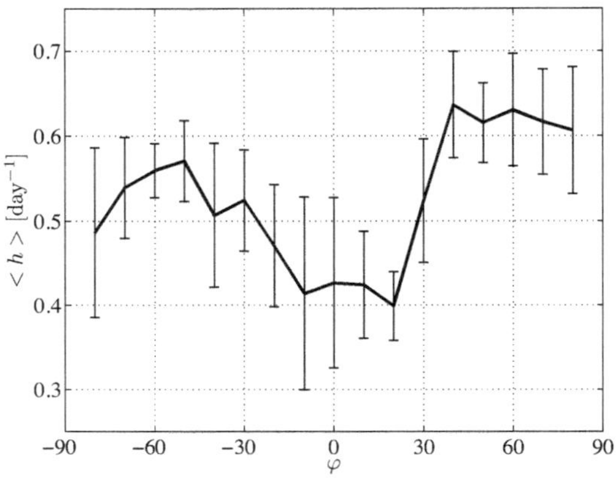

Figure 8. The zonal average of the topological entropy of Figure 7. The error bar denotes the standard deviation of the topological entropy values along a latitude.

It is worth noting that the atmosphere as a single system should be characterizable by a single value of the topological entropy. This is indeed valid for long term (of several months) observations. The 10-day-topological entropy used here is in fact a new location-dependent quantity, and it is a surprising empirical fact that a well-defined exponential scaling can be observed during such a short time. Anyhow, this local value of the topological entropy can be considered to be a new, useful measure of the chaoticity of the state of the atmosphere, and it can provide information on the speed of pollutant spread from a given location.

6 Escape Rate

6.1 General Concepts

Under certain circumstances chaotic behavior is of finite duration, i.e., the complexity and unpredictability of the motion can be observed over a *finite time interval* only. Nevertheless, there also exists in such cases a set in phase space responsible for chaos, which is, however, non-attracting. This type of chaos is called *transient chaos* and the non-attracting set is a *chaotic saddle* (for an introductory text see Tél and Gruiz, 2006). Since there are typically significant differences in the individual lifetimes, an *average* lifetime can be defined. To this end, it is worth following several motions instead of a single one: the study of particle ensembles is essential. To characterize the dynamics, one takes a preselected region, and starts $n_0 \gg 1$ trajectories in it. They escape the preselected region sooner or later, and the motion before escape appears to be chaotic. The number $n(t)$ of trajectories that do not leave the preselected region up to time t is thus a monotonically decreasing function of t. After a sufficiently long time (for t larger than some t_c), the decay in the number $n(t)$ of survivors is generally exponential:

$$n(t) \sim \exp(-\kappa t), \quad \text{for } t > t_c. \tag{11}$$

Coefficient κ is called the *escape rate* (Ott, 1993; Tél and Gruiz, 2006; Lai and Tél, 2011). Its reciprocal value can be considered as an estimate of the average lifetime of chaos. A nonzero escape rate is thus a new, important chaos characteristic: the larger the value of κ, the faster the escape/sedimentation process.

In the atmospheric context, the preselected region might be the entire atmosphere. The condition of escape is then the first arrival at the surface. It is interesting to see how turbulence and wet deposition influence the escape dynamics. We claim that the escape rates provide a kind of Lagrangian characterization of the entire deposition process.

6.2 Global Results

In order to determine global escape rates, we distribute $n_0 = 2.5 \times 10^5$ particles uniformly over the globe on different pressure levels on 1 January 2010. They are tracked in the ERA-Interim wind fields up to their escape, but at longest for 1 yr. To study the dependence of the escape rate on the particle size and on the initial altitude, simulations are run with radii of $r = 0, 1, 2, \ldots, 12$ µm and initial altitudes of $p_0 = 500, 700, 850$ and 900 hPa (Haszpra and Tél, 2013a). Note that the radius of a particle suddenly changes if the particle is captured by a raindrop, as discussed in the description of the RePLaT model (Section 2). (The limiting case of a "particle" with $r = 0$ µm can be considered as a gaseous contaminant in the atmosphere.)

To compare different effects, simulations are carried out in three setups that take into account:

1. advection, turbulent diffusion and precipitation,
2. advection and turbulent diffusion, and
3. only advection.

As a first example, Fig. 9 exhibits the number of survivors vs. time for a fixed initial altitude $p_0 = 500$ hPa (corresponding to free atmospheric initial conditions) for $r = 9$ µm particles in setup 1. As the aerosol particles are initially far from the surface, the curve starts with a plateau: no outfall from the atmosphere occurs within the first few days. After a short transition following the plateau (i.e. for $t > t_c \approx 1$–15 days for the different simulations), an approximately exponential decay can be seen for a few days (see the dashed line belonging to days 2–5 in Fig. 9). After some time, however, a crossover takes place and a slower exponential decay sets in for $t > 10$ days). Thus, we can speak of a short-term and a long-term exponential decay characterized by different exponents. The corresponding escape rates will be denoted by κ_s and κ_ℓ, respectively.

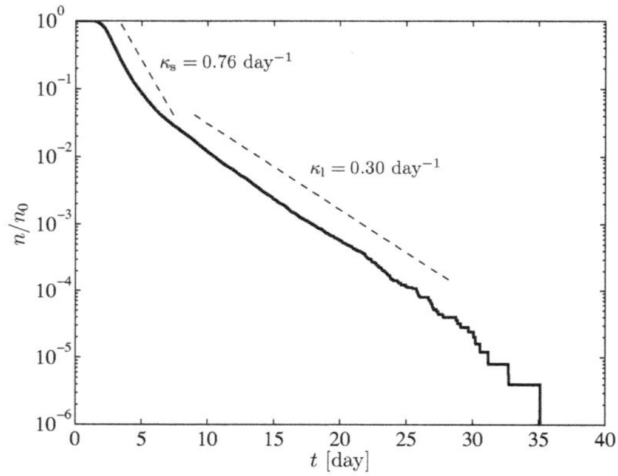

Figure 9. Proportion n/n_0 of the number of survivors in setup 1 as a function of time. $n_0 = 2.5 \times 10^5$ particles were distributed uniformly over the globe on $p_0 = 500$ hPa with $r = 9$ µm on 01.01.2010 00 UTC. Dashed lines illustrate the short-term and long-term decay processes.

Escape rates can be used as measures of the deposition process. Results obtained for different sizes show that both escape rates are at least 10 times larger for large aerosol particles (9 or 10 µm) than for small ones. The deposition process is thus very fast for large sizes. At any given size, the long-term escape rate is at least half or smaller than the short-term one. Since this difference appears in the exponent, we can safely speak about a separation of time scales in the deposition process.

Our findings illustrate that the naive expectation coming from dynamical systems theory according to which the global emptying is a random process described by a single exponential decay does not hold. In the atmosphere, instead, a short-term and a long-term dynamics can be identified, characterized by two different approximately exponential decays.

A detailed investigation shows that the long-term escape rate κ_ℓ computed for different initial pressure levels p_0 does *not* depend on the initial level in either setup (in contrast to κ_s which exhibits a strong p_0-dependence). The reason for this phenomenon might be the fact that particles surviving a long time in the atmosphere become well mixed. The independence of p_0 indicates that there exists a *global atmospheric chaotic saddle*, and the long-lived particles reflect properties of this set underlying the deposition dynamics. $\kappa_\ell(r)$ is thus a global atmospheric characteristic of particles of size r. The atmospheric saddle is likely to be time-dependent, and the $\kappa_\ell(r)$ values are characteristic of the time period investigated.

It is remarkable that κ_ℓ ranges over about two orders of magnitude although the radii vary over one decade only. The dependence is thus strongly nonlinear. The best approximate fit appears to be exponential

$$\kappa_\ell(r) \sim \exp(kr). \tag{12}$$

Exponent k is found to be $k \approx 0.33$–$0.38\ \mu\text{m}^{-1}$ for setups 1, 2, including rain and/or turbulent diffusion (see Fig. 10).

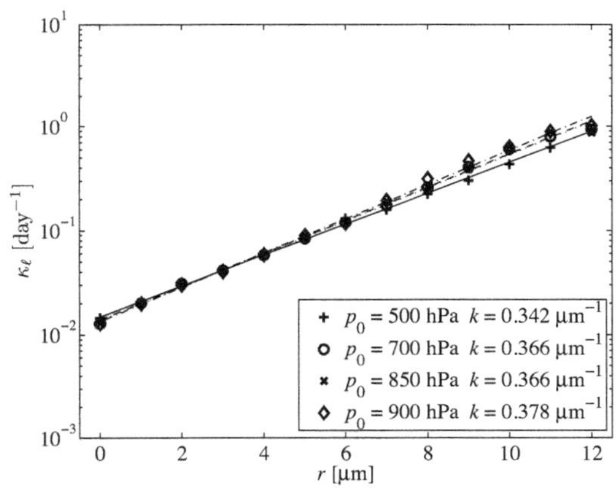

Figure 10. The dependence of the long-term escape rate on the size and initial altitude of the particles in setup 2 for different p_0 levels. Dashed lines indicate exponential fittings to κ_ℓ vs. r.

It is worth comparing the scaling of Eq. (12) with a naive estimate. The time needed to pass a fixed vertical distance Z with the terminal velocity (3) in non-moving air is Z/w_{term}. Since the terminal velocity is proportional to r^2, the time is proportional to r^{-2}. As the reciprocal of this time corresponds to the escape rate, the estimate results in a scaling proportional to r^2. The fit of this functional form to the data is much less satisfactory than that provided by (12). The difference between the power law behavior and the observed exponential one can only be interpreted by realizing that atmospheric winds play an essential role in the deposition process.

6.3 The Eruption of Mount Merapi

Mount Merapi in Indonesia had long-lasting eruption series in 2010, from late October to November. To study the outfall dynamics of aerosol particles, instead of the continuous eruptions, we simulate with RePLaT only a single volcanic ash puff of columnar shape of size $1° \times 1° \times 400$ hPa, centered at $\lambda_0 = 110.44°$ E, $\varphi_0 = 7.54°$ S and $p_0 = 500$ hPa (Haszpra and Tél, 2013a).

Figure 11 demonstrates the horizontal dispersion of the ash cloud containing $n_0 = 2.16 \times 10^5$ particles of $r = 5$ µm emitted at 00 UTC on 1 November in the ERA-interim wind field. As expected, such particles spread and reach very different regions in the atmosphere. Entering into different vertical levels, they become subject to different horizontal winds. The still strongly localized ash cloud on the 3rd day (Fig. 11, top left) spreads considerably up to the 7th day (Fig. 11, top right). It is worth mentioning that despite the simplifying one-puff assumption, this figure shows good agreement with the satellite image of sulfur dioxide tracers in the period of 4–8 November (http://earthobservatory.nasa.gov/NaturalHazards/view.php?id=46881). 20 days after the hypothetical emission, the particles initialized in a small volume cover a huge area and are well mixed in the midlatitudes of the Southern Hemisphere. Therefore the long-term escape rate found for this case ($\kappa' = 0.103$ µm^{-1}) is almost the same as the global escape rate κ_ℓ for $r = 5$ µm particles.

A remarkable feature of the bottom panel of Fig. 11, showing the ash cloud 20 days after the eruption, is that the distribution of the deposited (black) particles is fractal-like. There are large regions without any outfall, and the overall pattern is filamentary. The set of particles on the surface appears to trace out the intersection of the unstable manifold of the atmospheric chaotic saddle with the surface. This saddle might in principle be time-dependent, and what we see here is the set of these intersections in the interval November 7–20.

It is insightful to look at the vertical distribution of the particles over the time span followed. This can be seen in the form of a histogram in Fig. 12 for particles of $r = 10$ µm. The initially columnar shape is deformed into a Gaussian one that spreads as its center moves downwards. This behavior was also observed in a simple cloud model with aerosol particles (Drótos and Tél, 2011). It is remarkable, however, that after the center of the

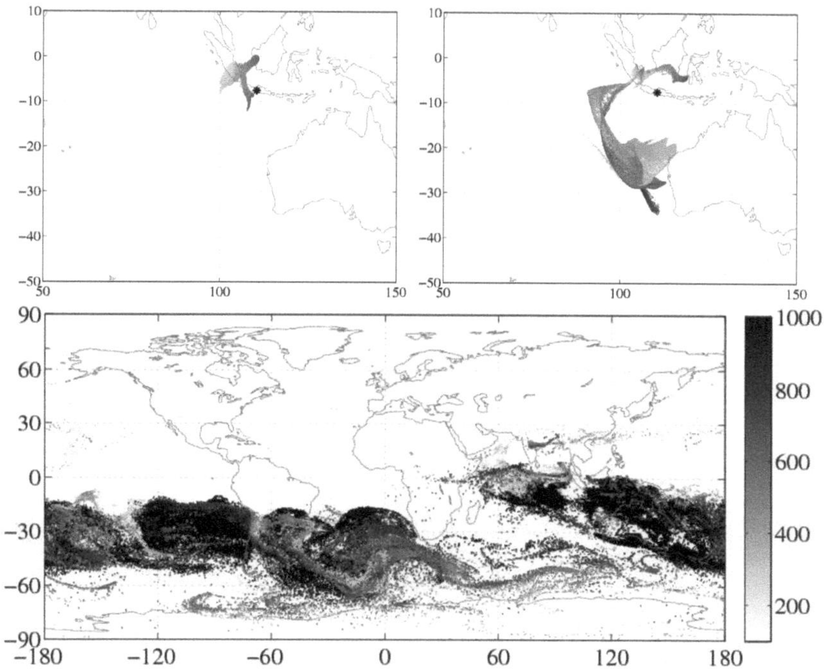

Figure 11. Dispersion of the volcanic ash consisting of $r = 5$ μm particles from the Mount Merapi eruption. Panels illustrate the geographical distribution 3, 7 and 20 days after the eruption taken place on 1 November, 2010, respectively. Graybar indicates the pressure level of the particles in hPa.

Gaussian distribution reaches the surface, and the majority of the particles is deposited, the small fraction of particles remaining aloft is distributed *widely* in the different layers. It is the fraction of these extreme survivors that is responsible for the second, long-term exponential decay observed. We believe that this wide altitudinal distribution of the extreme survivors is also the physical background of the time-scale separation described in the previous section.

6.4 Remarks

It is interesting to compare the escape process with precipitation activities. Only a small fraction of the $r = 10$ μm particles is found to leave the atmosphere in the first 6 days after the hypothetical eruption of Mount

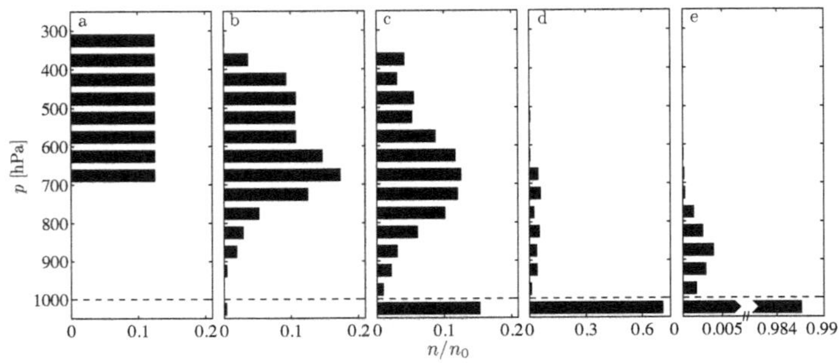

Figure 12. Vertical distribution of the proportion of particles with $r = 10\ \mu m$ in vertical layers of size 50 hPa for 0, 1, 3, 5, and 7 days after the hypothetical Mount Merapi eruption. The dashed horizontal line represents the surface.

Merapi. When they happen to reach the 850 hPa level precipitation starts playing an important role due to the frequent rainfall events above Indonesia. In the period of days 6–8 (6–8 November), particles reach a region of a cyclone with strong precipitation, therefore a large amount of particles are scavenged out by rain in this period. The particle distribution on the surface is found to strongly correlate with the rain intensity (Haszpra and Tél, 2013a). Indeed within 8 days, the majority of the particles falls out from the atmosphere in this case.

In summary, we have found that the emptying process of aerosol particles cannot be characterized by a single exponential decay. The global emptying process, from any height of the atmosphere, is governed by two temporal periods in which different exponential forms appear defining two different escape rates. The reciprocal value of the short-term escape rate is found to provide an estimate of the average residence time of typical particles. The analogous quantity belonging to the long-term escape rate characterizes exceptional particles that remain in the atmosphere for an extremely long time. It is interesting to note that the escape rates of particles of different sizes are found to vary in a broad range rather rapidly, roughly exponentially with the particle size. These investigations provide a Lagrangian foundation for the concept of deposition rates.

7 Ensemble Features and Outlook

It is important to note that there is always some uncertainty in the calculation of pollutant dispersion due to different error sources. One of these is the meteorological forecast produced by the numerical solution of the partial differential equations of the atmospheric hydrodynamics. Due to inaccuracies in the measurements and approximations used, the initial conditions of the meteorological model cannot be determined precisely. Initial errors are then amplified because of the sensitivity to initial conditions of atmospheric turbulence. Therefore, there is uncertainty in the meteorological forecasts which can be quantified by the so-called ensemble technique based on the execution of multiple meteorological simulations (Kalnay, 2003; Leutbecher and Palmer, 2008). These imply that there is a considerable difference in the driving provided by the velocity field $\mathbf{v}$ for the advection equations (2) or (4).

Dispersion models are usually run by a single forecast which is considered to be the best one. However, it can be useful to perform simulations using the whole ensemble forecast, i.e. producing an ensemble dispersion prediction in order to get a detailed and more reliable overview of the uncertainties and possible hazards related to the dispersion event.

Figure 13 shows the distribution of $r = 1$ µm particles in two ensemble members of an ECMWF ensemble forecast after 2.5 days tracking with the RePLaT model. The particles are initiated in a two-dimensional square of size $1^\circ \times 1^\circ$, centered at $\lambda = 141^\circ$E, $\varphi = 37.5^\circ$N and $p_0 = 500$ hPa at 00 UTC 12 March, 2011 (as a hypothetical emission well above the original emission height of the Fukushima Power Plant accident). For simplicity, we study only the effect of the variability in the wind field and do not take into account the impact of turbulent diffusion and precipitation (Haszpra et al., 2013; Haszpra and Horányi, 2014). Figure 13 demonstrates that even with this restriction strong dispersion variabilities may develop between the members both in the horizontal/vertical location and in the extension of the pollutant cloud in spite of the rather short time (2.5 days) passed after the emission.

Figure 14 provides a more systematic illustration of this statement. One can see that the horizontal locations of the center of mass of the pollutant clouds in the different ensemble members extend to a quite large area. The largest distance between the centers of mass is about 3375 km, on the order of half of the radius of the Earth. Also the standard deviations of the particles in the clouds, represented by the radii of the circles, vary in a wide range (from 35 km to 960 km). All this suggests that the problem of pollutant spreading should also be treated in the spirit of ensemble forecasts. The ensemble variability turns out to be rather strong and appropriate statistical measures should be introduced, some of which have been worked out in Haszpra et al. (2013).

In this paper we argued that a proper treatment of atmospheric material

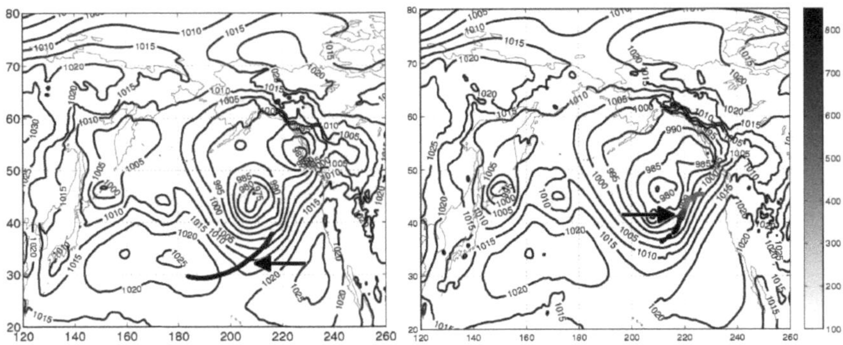

Figure 13. Geographical distribution of the pollutant cloud of $r = 1$ µm particles obtained from the 11th and 19th member of the ensemble forecast of ECMWF with RePLaT 2.5 days after the release at 00 UTC March 12, 2011 shown in the left and right panel, respectively. Graybar indicates the altitude of the particles in hPa. Contour lines represent the mean sea level pressure. The horizontal length of the pollutant cloud is about 2 times larger in the left than in the right panel, whereas the relation of the vertical extensions is approximately the opposite with a factor of about 3. The centers of mass are about 2000 km away from each other. The arrows point toward the pollutant clouds.

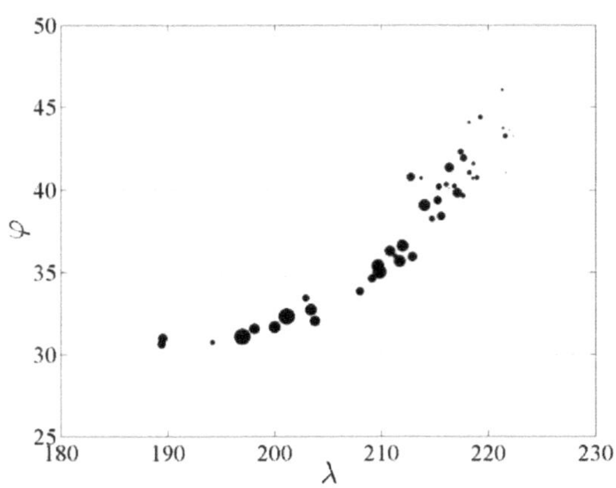

Figure 14. The horizontal geographical location of the center of mass of the pollutant clouds of $r = 1$ µm particles in the atmosphere for the 51 ensemble members after 2.5 days. The radii of the circles are proportional to the standard deviation of the clouds.

dispersion should be based on individual particle tracking using realistic size and density. The effect of turbulent diffusion and the scavenging of particles by precipitation can be incorporated as stochastic perturbations. This approach enables one to investigate the chaotic aspects of particle dispersion by means of dynamical systems concepts. The topological entropy and the escape rate were shown to be two useful measures not applied earlier in the atmospheric context. The uncertainty in the dispersion due to the unpredictability of the meteorological forecasts can also be studied.

All these features are, of course, subject to changes in the climatic conditions. For example, more intense local atmospheric circulation with more frequent and/or stronger cyclones might induce more stretched material lines generally and, along with this, larger 10-day-topological entropy values. In spite of these features, the global 10-day- and long-term topological entropies might become smaller than the corresponding values today due to the predicted (IPCC, 2013) decrease of the meridional temperature gradients in the future. The occurrence of high precipitation events might increase the proportion of the outfalling particles locally, while regions with less precipitation and possibly stronger updrafts than during present climate conditions can decrease the deposition. The overall effect to the value of the escape rate would require additional numerical simulations. More intense local circulation might also enhance the uncertainty of meteorological forecasts and, therefore, it might increase the variability of the ensemble dispersion simulations based on these forecasts. The detailed investigation of these issues remains a task for the future.

Acknowledgements

The authors would like to thank G. Drótos, Z. Ferenczi, V. Homonnai, A. Horányi, I. Ihász, L. Lagzi and G. Wotawa for useful comments and remarks. This work was supported by the Hungarian Science Foundation under Grant no. OTKA NK100296 and by the von Humboldt Foundation.

Bibliography

H. Aref. Stirring by chaotic advection. *Journal of Fluid Mechanics*, 143(1): 1–21, 1984.

D. P. Dee et al. The ERA-Interim reanalysis: Configuration and performance of the data assimilation system. *Quarterly Journal of the Royal Meteorological Society*, 137(656):553–597, 2011.

R. R. Draxler and G. D. Hess. *Description of the HYSPLIT_4 modeling system.* Air Resources Laboratory, Silver Spring, Maryland, 2004. NOAA

Technical Memorandum ERL ARL-224.

G. Drótos and T. Tél. Chaotic saddles in a gravitational field: The case of inertial particles in finite domains. *Physical Review E*, 83(5):056203, 2011.

A. J. Dyer. A review of flux-profile relationships. *Boundary-Layer Meteorology*, 7:363–372, 1974.

T. Haszpra and A. Horányi. Some aspects of the impact of meteorological forecast uncertainties on environmental dispersion prediction. *Időjárás*, 118(4):335–347, 2014.

T. Haszpra and T. Tél. Volcanic ash in the free atmosphere: A dynamical systems approach. *Journal of Physics: Conference Series*, 333:012008, 2011.

T. Haszpra and T. Tél. Escape rate: a Lagrangian measure of particle deposition from the atmosphere. *Nonlinear Processes in Geophysics*, 20 (5):867–881, 2013a. doi: 10.5194/npg-20-867-2013.

T. Haszpra and T. Tél. Topological entropy: a Lagrangian measure of the state of the free atmosphere. *Journal of the Atmospheric Sciences*, 70 (12):4030–4040, 2013b. doi: 10.1175/JAS-D-13-069.1.

T. Haszpra, I. Lagzi, and T. Tél. Dispersion of aerosol particles in the free atmosphere using ensemble forecasts. *Nonlinear Processes in Geophysics*, 20(5):759–770, 2013. doi: 10.5194/npg-20-759-2013.

J. L. Heffter and B. J. B. Stunder. Volcanic Ash Forecast Transport And Dispersion (VAFTAD) model. *Weather and Forecasting*, 8(4):533–541, 1993.

IPCC(2013). *Climate Change 2013: The Physical Science Basis, Working Group I Contribution to the Fifth Assessment Report of the Intergovernmental Panel on Climate Change.* Cambridge University Press, New York, 2013.

B. Johnson et al. In situ observations of volcanic ash clouds from the faam aircraft during the eruption of eyjafjallajkull in 2010. *Journal of Geophysical Research: Atmospheres*, 117(D20), 2012. ISSN 2156-2202. doi: 10.1029/2011JD016760.

E. Kalnay. *Atmospheric modeling, data assimilation and predictability.* Cambridge University Press, Cambridge, 2003. ISBN 0-521-79179-0.

Y.-C. Lai and T. Tél. *Transient Chaos: Complex Dynamics on Finite Time Scales.* Springer, New York, 2011. ISBN 978-1-4419-6986-6.

M. Leutbecher and T. N. Palmer. Ensemble forecasting. *Journal of Computational Physics*, 227:3515–3539, 2008.

S. Newhouse and T. Pignataro. On the estimation of topological entropy. *Journal of Statistical Physics*, 72:1331–1351, 1993.

E. Ott. *Chaos in Dynamical Systems.* Cambridge, 1993.

H. R. Pruppacher and J. D. Klett. *Microphysics of Clouds and Precipitation.* Kluwer Academic Publishers, Dordrecht, Netherlands, 1997. ISBN 978-0-7923-4211-9.

C. Searcy, K. Dean, and W. Stinger. PUFF: A high-resolution volcanic ash tracking model. *Journal of Volcanology and Geothermal Research*, 80: 1–16, 1998.

J. H. Seinfeld and S. N. Pandis. *Atmospheric Chemistry and Physics: From Air Pollution to Climate Change.* John Wiley & Sons, Inc., 1998. ISBN 0-471-17815-2.

B. Sportisse. A review of parameterizations for modelling dry deposition and scavenging of radionuclides. *Atmospheric Environment*, 41(13):2683–2698, 2007.

A. Stohl et al. *The Lagrangian particle dispersion model FLEXPART version 8.2*, 2010. Technical report.

A. Stohl et al. Xenon-133 and caesium-137 releases into the atmosphere from the Fukushima Dai-ichi nuclear power plant: determination of the source term, atmospheric dispersion, and deposition. *Atmospheric Chemistry and Physics*, 12(5):2313–2343, 2012.

T. Tél and M. Gruiz. *Chaotic Dynamics: An Introduction Based on Classical Mechanics.* Cambridge University Press, 2006. ISBN 978-0-512-83912-9.

I. B. Troen and L. Mahrt. A simple model of the atmospheric boundary layer; sensitivity to surface evaporation. *Boundary-Layer Meteorology*, 37:129–148, 1986.

The parameter optimization problem in state-of-the-art climate models and network analysis for systematic data mining in model intercomparison projects.

Annalisa Bracco[*][*] Richard K. Archibald[†], Constantine Dovrolis[‡], Ilias Foundalis[‡], Hao Luo[*] and J. David Neelin[§],

[*] School of Earth and Atmospheric Sciences, Georgia Institute of Technology, Atlanta, GA, USA

[†] Climate Change Science Institute, Oak Ridge National Laboratory, Oak Ridge, TN, USA

[‡] College of Computing, Georgia Institute of Technology, Atlanta, GA, USA

[§] Dept. of Atmospheric and Oceanic Sciences, UCLA, Los Angeles, USA

Abstract The focus of this work is on two major problems facing the scientific community when using increasingly complicated climate model outputs to investigate the past and future evolution of our climate. On one hand, it is important to assess the reliability of such models and how their response to increased greenhouse gas concentrations may depend on the parameters and parameterizations chosen; on the other, it is fundamental to improve our ability to validate and compare model results in a robust, compact, and meaningful way. Understanding how sensitive climate models are to changes in their parameters is of fundamental importance when addressing the problem of modeled climate sensitivity. Here a quadratic metamodel that uses a polynomial approximation to describe the parameter dependency is presented together with its application to the Community Atmospheric Model, CAM, in its two latest versions. Furthermore, the application of complex network analysis to climate fields is briefly summarized and a novel methodology that allows for robust model intercomparisons is presented together with a set of metrics to quantify the topological properties of model outputs. The application of the network analysis to outputs from the Coupled Model Intercomparison Project - Phase 5 (CMIP5) completes the notes.

[*]The authors wish to thank the generous support of the US Department of Energy through the SciDAC program, DE-SC0007143, and of the National Science Foundation, grant DMS 1049095 that supported this work.

A. Provenzale et al. (Eds.), *The Fluid Dynamics of Climate*,
CISM International Centre for Mechanical Sciences
DOI 10.1007/ 978-3-7091-1893-1_5

1 Introduction

General circulation models currently used for understanding current and past climate and for predicting its evolution in the future, exhibit substantial spreads in their equilibrium sensitivity, implying that the magnitude of their temperature increase in response to a doubling carbon dioxide is uncertain. The mean temperature increase over the 21st century projected by models in the last Intergovernmental Panel on Climate Change assessment continues to resist any narrowing of the range of estimates even in the historical integrations (1; 2), while the evolution of the major modes of variability of the climate system diverges (3; 4). Large uncertainties for end-of-century climatic variables prevails not only in the simulation of future surface-air temperatures, but also in precipitation (5), cloud cover (6; 7), winds (8), sea level (9), sea ice (10), and other variables of importance for socio-economic, ecological and human-health impacts. It is fair to state that while no legitimate doubts exist about the future rise in global temperatures and about additional changes in climate being significant, many questions remain about the extent of the changes, not only in the mean, but also and even more so in the variability of climatic fields, in space and time.

Despite successful representation of large-scale averages, precipitation, cloud properties and distributions, water content and paths, and cloud radiative effects (11; 12; 13) prove difficult to constrain toward observations at regional scales. Consequently, the confidence in regional-scale projections of future changes in the mean state and in major modes of variability (3) is hampered. The challenges faced by modelers in trying to reduce this uncertainty include the multiplicity and nonlinearity of the processes and feedbacks that the climate system contains, its high-dimensionality, and the computational requirements (15). A common experience for modelers is that the simulated climatology and/or variability exhibit high sensitivity to parameterization changes and parameter choices. This is especially true when changes are associated with the microphysics of cloud formation and convection, aerosol emissions and processes, or ocean mixing, as nicely shown in other contributions to this book. The end result is that for any such change we are faced with improvements in certain variables or geographical regions, and degradation in others.

New approaches to quantify and characterize uncertainties in climate model simulations have been developed in recent years, as briefly summarized later on. Here we focus on two contributions to the investigation of parameter sensitivity and model uncertainties developed by the authors. Specifically, we present a multi-objective approach and a metamodel as a strategy for fitting parameter dependence (15; 16), and a fast, scalable and cutting-edge computational toolbox based on complex network analysis to investigate local and non-local statistical interrelationships in climate model outputs (14).

2 Multiobjective optimization to understand parameter model sensitivity

Diagnosing uncertainties associated with parameterizations and parameter choices in climate models is a key challenge for the science community, and one that is becoming more pressing the larger is the parameter space to be explored. Any new release of a general circulation model indeed represents both an increase in the degrees of freedom and parameter choices, and a significant departure from previous versions. For example, the US climate research community recently moved from the fourth version of the Community Atmospheric Model (CAM4) to its fifth (CAM5). Compared to its previous version, CAM5 is characterized by a new microphysics parameterizations (17) and the representation of cloud processes in CAM5 differs significantly from that in CAM4. Changes in the cloud physics package of CAM5 include a new shallow convection scheme (18) that uses a realistic plume dilution equation and closure, and aims at a more accurate simulation of spatial distribution of shallow convection; a new parameterization of microphysical processes (19) based on a prognostic, two-moment formulation for cloud droplet and cloud ice, liquid mass and number concentration; a new macrophysics scheme (20) that imposes consistency between cloud fraction and cloud condensate; a new moist turbulence scheme (21) that allows for the treatment of stratus-radiation-turbulence interactions and is based on the parameterization of eddy diffusivity as function of turbulent kinetic energy, entrainment rate and a stability function, and a new radiation scheme (22) that includes an efficient and more accurate correlated-k method for evaluating radiative fluxes and heating rates. Specific to the cloud/transport scheme in the microphysics package, a unified PDF-based cloud scheme is introduced (23). The mass-bases bulk microphysics scheme of CAM4 is therefore substituted by a two-moment scheme (one for mass and one for number concentration) that implements an analytical representation of the size distribution of droplets and uses the moments of the distribution. Overall, those changes have improved the cloud representation of CAM5 when compared to CAM4 (11). CAM5 reduces known biases such as the underestimation of total cloud and the overestimation of optically thick cloud, and with its radiatively active snow ameliorates the underestimation of midlevel cloud. The number of degrees of freedom in the parameter space of CAM5 however, is larger than in CAM4, complicating to a greater degree the tuning process if done by brute force. Tuning by brute force refers to the retesting and tuning of an optimized set of parameters optimized according to the modeler needs - every time a given parameter value or parameterization scheme are modified.

Similarly substantial changes affected several other models, as in the case of the Institute Pierre Simon Laplace (IPSL) climate model that in 2011 underwent a recasting of the parameterization of turbulence, convection and clouds (24), or of the Earth system model of Max Planck Institute for Meteorology (MPI-ESM) that in its latest version added a direct representation of the carbon cycle, modified the representation of the middle atmosphere, of shortwave radiative transfer, surface albedo and aerosol, and implemented a land surface module with interactive vegetation dynamics (all changes have been documented through a special electronic edition of the Journal of Advances in Modeling Earth Systems published in 2013 and can be found at http://www.mpimet.mpg.de/en/science/models/mpi-esm/james-special-issue.html).

The diagnosing of uncertainties associated with parameterizations and parameter choices in climate models has been attempted with various methodologies. A recent example of brute force exploration combined to a stochastic importance-sampling algorithm that allows for progressive convergence to optimal parameter values is described in (25) for CAM5. Alternatively, a downhill simplex method can be used to tune and improve the climatology of a coupled model, as suggested by (26). Or the so-called perturbed physics approach that consists in obtaining large ensembles of model runs by perturbing poorly constrained parameters to account for the incomplete or imprecise knowledge of their actual values can be applied following (27). Examples of its application are found in (28; 29) and (30). (31) proposed Bayesian inference together with a stochastic sampling algorithm to estimate the posterior joint probability distribution for given, uncertain parameter sets given a prior probability for selecting reasonable values for each set, and (32) introduced the idea of surrogate-based optimization, where a computationally cheap and yet reasonably accurate model, build and updated using the output from any state-of-the-art GCM, replaces the more complex one in the optimization process to obtain a model optimum.

Here we discuss in some detail the multiobjective optimization methodology proposed by (15) and (16). It is a computationally efficient framework for the systematic investigation of parameter space in climate models that consists on approximating the models parameter dependence to a low-order polynomial. More importantly, it presents the advantage of requiring a limited number of model integrations to explore the model sensitivity.

In essence, the multiobjective optimization represents a strategy to explore parameter dependencies and model performances for climatologies and mean state changes. The optimization can be performed repeatedly for as many objective functions as the user desires, and allows for investigating and interpreting the dependence of the model solution on the simultaneous change of multiple parameters. In most cases it is sufficient to run the model at the standard parameter value, and its minimum and maximum reasonable values (i.e. the minimum and maximum acceptable on the base of physical or chemical constrains) to reconstruct global averages and/or regional patterns for entire plausible range. The approach stems from the engineering and theoretical optimization literature, and assumes that the error metric varies smoothly whenever parameters are changed. A smooth response to parameter changes is not an a priori property of general circulation models. However, a theoretical argument to justify linear response theory in climate science has been proposed recently by (33). The smoothness assumption has been verified by an extensive suite of experiments performed using the ICTP-AGCM (International Center for Theoretical Physics - Atmospheric General Circulation Model) (15; 16), by the non-hydrostatic regional simulations presented in (34), and by explorations performed using CAM4 (35) and CAM5 of which examples are given below.

The multiobjective optimization methodology builds upon the general smoothness of the response of climate models to changes of most parameters, and allows to objectively assess regional tradeoffs and optima at low computational cost, aiding sensitivity studies. A climate field of interest, $\phi(\mathbf{x}, t)$, for example a climatology or a regression on a particular index, can be expressed as

$$\phi_{mm} = \phi_{std} + \sum_{i=1}^{N} a_i \mu_i + \sum_{i=1}^{N} \sum_{j=1}^{N} b_{i,j} \mu_i \mu_j \quad , \tag{1}$$

where $\mu_i = \mu_{i_{pert}} - \mu_{i_{std}}$ is the parameter i taken relative to its standard value $\mu_{i_{std}}$, N the number of parameters considered, $a_i(\mathbf{x}, t)$ is a high-dimensional vector containing the linear coefficients for each parameter at each grid point in time, and $b_{i,j}(\mathbf{x}, t)$ represents the quadratic (diagonal) and interaction (off-diagonal) terms, assuming $b_{i,j}(\mathbf{x}, t) = b_{j,i}(\mathbf{x}, t)$. Thus a fit procedure of order N allows to estimate the linear sensitivity and the quadratic nonlinearity, while the off-diagonal coefficients, obtained with a number of simulations of order N^2, can be calculated from the corners of pairwise planes. An in-depth discussion of the computation cost and advantages of this approach is provided in the Appendix to (15). Here we stress that the number of parameter points required is at most $O(N^2)$, which is by far less costly at large N than the $O(d^N)$ runs required by brute-force sampling at a given density d. For example, assuming $d = 3$, and 20 parameters, the multiobjective optimization demands a minimum number of integrations equal to $2N + N(N-1)/2 = 230$ plus the standard run, augmented by the verification points (in our experience usually of order 40%), to be compared to $3x10^9$.

With relevance to the CAM model, we have been extending the uncertainty quantification using the multiobjective optimization to test how changes in several parameters modify the performances of the AGCM in both its 4 and 5 versions. As an example, we consider the model error dependency for changes of the critical relative humidity threshold for low cloud formation, a parameter indicated with RHMINL. Although this parameter is consistent in the two CAM versions, and so it is the standard value recommended for use (0.90), the distributions for most fields are different in the two models as a result of the different parameterizations adopted. Five 50-year long runs are performed with each version of CAM increasing RHMINL from 0.85 to 0.95 at equally spaced intervals. In all cases CAM is forced using monthly varying sea and land surface temperature climatologies build using reanalysis data over the 1979-2008 period (monthly data are averaged over the 30 year period to build the climatological annual cycle used to force CAM at its lower boundary). Figure 1 presents the globally averaged root-mean-square (RMS) error of the surface stress exerted by the wind to the Earth surface and of the geopotential height field at 500 hPa (Z500) in boreal summer (June to August, JJA) relative to the National Centers for Environmental Prediction (NCEP) reanalysis (36). The plots display the RMS error for each of the 5 simulations, and the fit obtained applying the metamodel in Eq. 1 using only the linear (green line) or linear and quadratic (red line) coefficients and the model output at the standard value and at the minimum and maximum explored. The linear coefficients are sufficient to capture the general behavior of the model dependency for Z500, but not for the wind stress, particularly in CAM5 where the model dependency follows a convex trajectory in the global RMS. CAM5 displays a reduction in the global RMS error for the surface variable compared to CAM4, but no significant improvement is found in the representation of geopotential height. Figure 1 highlights also the existence of optima at the limit of the permissible parameter range for some variable (here for both variables in CAM5), indicating that the parameterization for the low level cloud formation, unsurprisingly, still warrants close attention. Finally, and more importantly, the dependence for varying RHMINL in the two model versions is opposite for the two chosen variables. In Z500 the RMS of the global error increases monotonically for increasing value of RHMINL in CAM5, while is at its maximum at RHMINL = 0.85 and decreases for

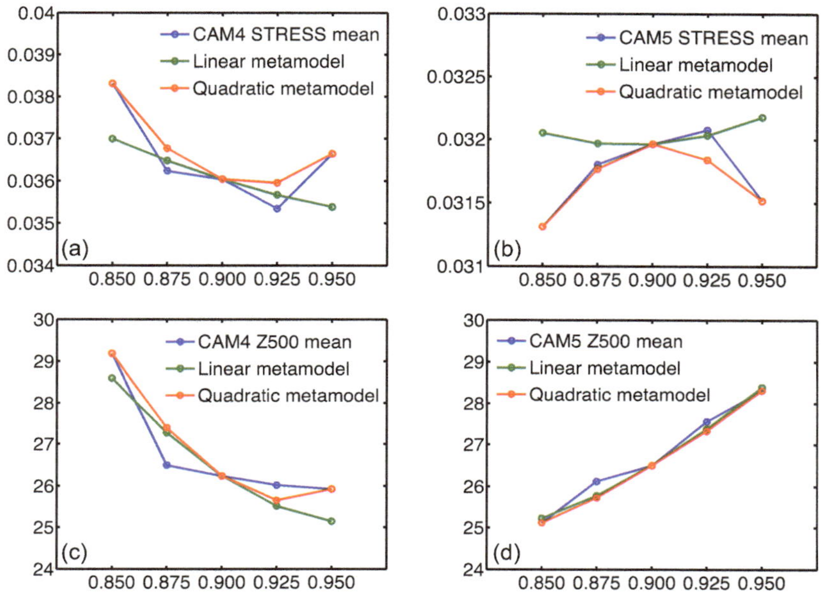

Figure 1. Root-Mean-Square (RMS) error of the CAM4 (left) and CAM5 (right) climatology of (a-b) near-surface wind stress, and (b-c) 500 hPa geopotential height (Z500) for varying RHMINL in JuneAugust (JJA) relative to the National Centers for Environmental Prediction (NCEP) reanalysis. The CAM values (blue) are compared to the linear (green) and quadratic (red) metamodel reconstruction based on the endpoints and standard value for RHMINL. By construction the linear metamodel gives quadratic terms with positive curvature in the RMS error. Units are Pa for wind stress, and m for Z500.

increasing parameter value achieving its minimum at 0.925 in CAM4. In wind stress largest and smallest RMS errors are found in proximity of the standard value in CAM5 and CAM4, respectively.

Figure 2 shows the comparison between the distribution of the RMS error in Z500 in boreal summer with respect to the NCEP reanalysis at RHMINL = 0.925 in CAM and as reconstructed by the metamodel. The error is concentrated at latitudes greater than 50o in both hemispheres, is always positive (indicating a model underestimation of the observed patterns common to other models, see for example (37) in the northern portion, negative between 40° and 60^oS, and positive again over Antarctica. At the given parameter value, CAM4 performs better in the northern hemisphere than CAM5 but the quadratic metamodel underestimates its error by approximately 10% - 25% of its RMS value. This suggests that the general behavior and patterns are well captured, and this is usually sufficient for an investigation that aims at finding a good compromise in the parameter settings; on the other hand the nonlinearities may contribute more than a quadratic corrections and further polynomial terms may be required if indeed the exact value of error magnitude is a modeler priority. For CAM5, the RMS error maps reveal better agreement with the quadratic metamodel reconstruction but an overall deterioration of the Z500 climatology over most of the northern hemisphere compared to the previous version of the model.

3 Network analysis to quantify climate interactions

The fast growing availability of observations from remote measuring platforms such as satellite and radars, as well as the increasingly more detailed outputs from global-scale climate models, contribute a continuous flow of terabytes of spatiotemporal data. The last two decades have been characterized by a rate of data generation and storage that far exceeds the rate of data analyses. While the literature in statistical analysis applied to climate fields, observed or modeled, is mature, systematic efforts in climate data mining are still lacking. Evaluating climate model outputs in a fast, scalable, and robust way while condensing information and allowing for meaningful comparisons is therefore one of priorities of the scientific community. In the last decade the application of network analysis to climate science have received some attention, beginning with the seminal paper by (38). In computer science, complex network analysis refers to a powerful tool used to investigate local and non-local statistical interrelationships. Such tool is composed by a set of metrics, models and algorithms commonly used in the study of complex nonlinear dynamical systems, and its main premise is that the underlying topology or network structure of a system has a strong impact on its dynamics and evolution (39).

Since 2004 climate networks have been used to investigate climate shifts relating network changes to El Niño Southern Oscillation (ENSO) activity (40; 41; 42; 43), identify global scale structures responsible for energy transfer through the ocean (44), evaluate climate models and identify teleconnections (45; 46), and represent the interaction between different climate variables as a network (47). In most cases, edges between nodes of the climate network are inferred using linear or non-linear similarity measures (for

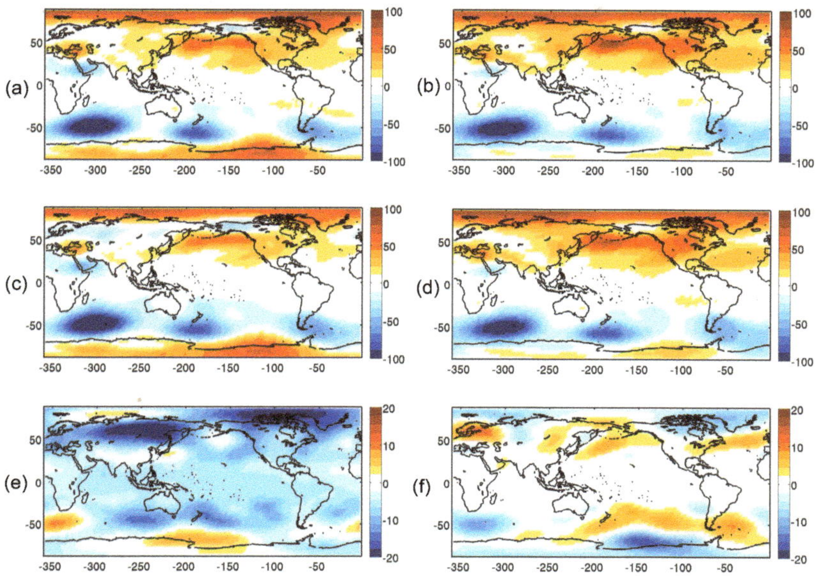

Figure 2. (a-b) Spatial distribution of Z500 RMS error relative to NCEP reanalysis for RHMINL = 0.925 and all other parameters at their standard values in JJA. (c-d) RMS error reconstructed using the quadratic metamodel. (e-f) Difference between model and metamodel reconstructed error (rescaled for clarity). Left: CAM4; Right: CAM5. Unit: m.

example Pearson correlation, mutual information, or phase synchronization) (48; 49), and the network is constructed as a (weighted or binary) undirected graph. It is well noted that correlation does not imply causation (50), and the next challenge in climate network analysis is arguably to move from undirected correlation based networks to directed causal ones to be able to identify feedback loops between the different variables of the climate system. Additionally, the network inference methods adopted in the works just cited construct graphs in which two cells are not considered connected and they are consequently pruned - or their correlations are set to zero - whenever the cross-correlation between them is less than a given threshold. Cell-level pruning makes the network inference process less robust and limits the reliability of model intercomparison exercises that implement it.

To overcome these limitations, to quantify differences and analogies between models or modeled and observed quantities, and to extend the application of network analysis to model ranking and intercomparison, we have developed a novel, fast, robust and scalable methodology to examine, quantify, and visualize climate patterns and their relationships (14). It is based on a two-layer network representation. At the first layer, gridded climate data are used to identify areas, i.e., geographical regions that are highly homogeneous in terms of the given climate variable, and that practically correspond to known modes of climate variability. At the second layer, the identified areas are interconnected with links or connections of varying strength, forming a global climate network.

The network inference we proposed is a three-step process. First we construct a "cell-level network"; second we apply a clustering algorithm to identify the nodes or areas, i.e. non-overlapping geographically connected regions that are homogeneous to the underlying variable; third we compute weighted links between areas to assess their connections. The cell-level network is constructed computing the Pearson cross-correlation between the detrended time series of the climate variable of interest for all grid cells pairs. Quite naturally time lags can also be taken into account in the cross-correlation calculation to build a dynamical network. All pair correlations are retained and the resulting cell-level network is a complete weighted graph (i.e. a link exists between all pairs of grid cells). This characteristic differentiates our method from most prior work on climate networks where a threshold to prune non-significant correlations is applied (44; 49; 51; 41), and ensures robustness of the area-level structure, allowing for reliable comparison of different networks, as extensively tested in (14). The clustering algorithm relies on a single parameter, τ, that varies between models or datasets considered and controls the homogeneity of areas to the underlying climate variable. τ represents the minimum average pair-wise correlation between cells of the same area at a given significance level. The algorithm aims also to minimize the number of areas identified; the problem is shown to be NP-Complete (14), thus the algorithm must rely on greedy heuristics. Finally, links are computed from the area cumulative anomalies weighted by the cell sizes. The weighted link between two areas is equal to the covariance between the corresponding cumulative anomalies; links - positive or negative - are computed for all pairs of areas to obtain a complete weighted graph. Link maps allow the visualization of the (weighted) connections between any given area and all others in the network. Areas are also characterized by their weighted degree or strength, defined as the sum of the absolute link weights. Strongest areas exert the greatest impact

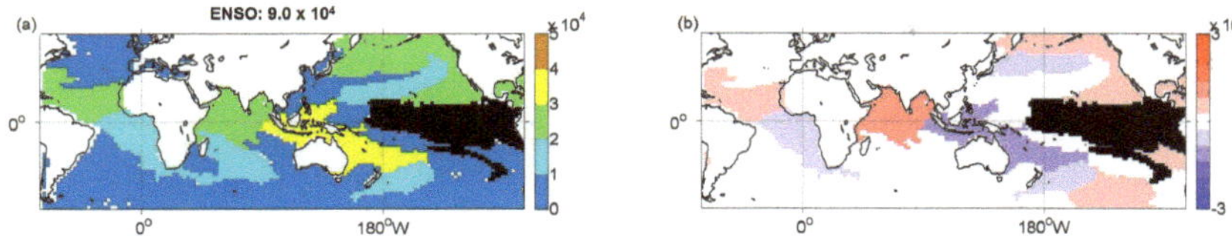

Figure 3. Strength (a) and ENSO-related link (b) maps for the networks calculated using the HadISST during the period 1956-2005 in boreal summer (JJA). The strength of the ENSO-related area exceeds the colorscale and is saturated. Its value is indicated at the top of the stregnth panel.

on climate variability.

Using complex network analysis to evaluate models' performance and their dependencies yields several desirable properties. The investigation is not locked into a particular climate mode or index - or to a set of indices - from the outset. From a set of climate model runs, different users can evaluate networks for various fields and/or regions, and derive model-dependent areas and their links in lieu of climate modes and their teleconnections. The methodology is scalable, and allows for direct, robust comparisons between different models or the same model integrated using different parameters, parameterizations, or forcings. Furthermore, it is immediate to include an estimate of internal variability when multiple ensemble members are available, which can be directly compared to contributions from different forcings, and to model trajectories over time.

An example of strength and link maps is provided in Figure 3 for the Hadley Center sea surface temperature (HadISST) reanalysis (52) shown here for boreal summer (JJA) over 1956-2005. The strongest area identified in the network corresponds to ENSO and it is linked to the Indian Ocean where SSTs are found to be warmer than average in correspondence of El Niño events, and vice versa for La Niñas.

To quantify similarities and differences between two networks in a compact way, we developed a new metric and adopted one from the complex network literature. We consider networks N and N', for the same variable (for example sea surface temperatures for a realizations of the Community Climate System Model Version 4, CCSM4 (53), that uses CAM4 in its atmospheric component, and for HadISST), each of size n grid cells. First, we compare their strengths by defining a network distance D as

$$D(N, N') = \frac{\sum_{i=1}^{n} |W_N(i) - W_{N'}(i)|}{\sum_{i=1}^{n} |W_N(i) - \hat{W}_{N'}(i)|} \,. \tag{2}$$

where $W_N(i)$ is the weight assigned to i−th grid cell in network N, and is equal to the strength of the area to which cell i belongs and $\hat{W}_{N'}$ is the strength of a randomly chosen grid cell in network N'. The normalization accounts for small distances in the nominator of D whenever area strengths have narrow distributions. The smaller the distance, the more similar two networks are in their strength distribution. Second, we measure the spatial likeness of the areas in the two networks by the Adjusted Rand Index (ARI) (54; 55). Any pair of cells that belong to the same area in both N and N', or that belong to different areas in both networks, contributes positively to the ARI. Conversely, any pair of cells that belong to a given area in one partition but to different areas in the other, contributes negatively. The ARI ranges between 0 and 1, with 1 denoting perfect similarity. The metric is adjusted to ensure that the distance between two random partitions is zero. ARI and D can be considered globally i.e. spaced averaged over the whole model domain - for example to compare the evolution of the network for a specific field under increases greenhouse gas forcing, or regionally i.e. spaced averaged over a specific region -, to analyze the response of a limited number of areas, for example to focus on changes in precipitation over Asia whenever aerosol effects are excluded or incorporated in a model projection.

A global application is presented in Figure 4, where the time evolution of D and ARI for the sea surface temperature field in JJA for five models that participated to the Coupled Model Intercomparison Project phase 5 (56) is shown. The chosen models are CCSM4, MPI-ESM, IPSL, HadGEM2 (Hadley Global Environment Model 2, (57)) and GISS-E2H (Goddard Institute for Space Studies model E2H distribution, (58). The top panel displays ARI vs D for three or four ensemble members in their historical period 1956-2005 calculated with respect to the HadISST reanalysis over the same time frame. The two metrics are also evaluated for two other SST reanalysis data-sets, ERSST-V3 (59) and SODA 2.1.6 (60) again with respect to HadISST to provide contest for the comparison. The middle panel shows ARI vs D for the same integrations projected into the near future (2051-2100), and the bottom panel presents the evolution of both metric for the only available integration projected to 2300 following the highest Representative and Extended Concentration Pathways, RCP8.5 and ECP8.5 (61). D and ARI in the middle and bottom panels are calculated with respect to their historical counterpart and therefore quantify the differences between present and future climate modes of variability and their links. Both metrics are also mapped to the amount of white Gaussian noise (WGN) that added to the climate field with network N will result in a network N such that $ARI(N, N') = ARI(N, N")$ and $D(N, N') = D(N, N")$. Over the historical period two models, CCSM4 and HadGEM2, outperform the remaining three at least for some ensemble member. CCSM4 is characterized by an internal variability, measured by the intra-ensemble spread, larger than any other model, with one member largely underestimating the strength of ENSO and its teleconnections (not shown) and therefore being penalized in the evaluation of D. In the RCP8.5 scenario changes in the network properties between the second half of the twentieth and twenty-first centuries are modest for most model members (3), and contained within the spread between different SST reanalyses in the historical period, despite substantial trends. The HadGEM2 and CCSM4 members that do not follow this behavior are characterized by a general weakening of all areas and in particularly of the ENSO related one, while the MPI-ESM and IPSL runs with the greater distance from historical display a strengthening of the ENSO and Southern Ocean area, respectively (not shown; see maps for the boreal winter season in (3). After 2100, all models display significant changes in the strength and, with the exception of IPSL, in the shape and size of major areas. Three models reduce the strength and size of the ENSO node, and evolve towards weakening all tropical areas and their links over the 23rd century. Figure 5 provides an example of the drastic reduction in area size and connectivity that characterizes most models by displaying the strength and link maps over 1956-2005 and 2251-2300 for HadGEM2. The IPSL network changes in the tropical connectivity (especially over the Indian Ocean, which is not linked to ENSO) but maintains areas and extratropical links. Finally, the MPI-ESM SST network strengthens slightly over time (Figure 5, left panels), and better compare to the HadISST for the shape of the major areas. All historical links from the ENSO related area are reproduced in the future, with the exception of the connection with the South Tropical Atlantic, positively correlated to ENSO in the model and negatively in the observations. In (3) we conclude that the uncertainty in the projected connectivity after 2100 in many regions exceeds the uncertainty associated with the equilibrium sensitivity.

3.1 Conclusions

We have presented results summarizing recent work on the parameter sensitivity of climate models showing that a quadratic metamodel for spatial, seasonal fields permits reconstruction of multiple objective functions of interest at a reduced computational cost compared to existing practices.

The metamodel is simple but very flexible, allowing for the evaluation of a large number of variables, regions, and parameter combinations with a limited number of integrations, and provides a reliable estimate of the spatial distribution of model biases. Solutions at the boundary of the admissible parameter range, i.e. boundary optima, are common to climate models as shown for CAM4 and CAM5 here and for the ICTP-AGCM in (15) and (16), and point to the parameterizations that need close scrutiny, as in the case of convection and cloud microphysics.

We have also introduced few concepts on network analysis and reviewed some of the most recent applications to climate science. Our work has focused on developing a methodology to capture major climate modes and their connectivity while allowing for a robust comparison between different model outputs. Focusing on ensembles from 5 coupled climate models in the CMIP5 catalog, we have shown that according to our analysis most models respond to increasing emissions and warming by changing only slightly their climate modes until the end of this century. Consequently the spread between detrended scenarios measured in terms of ARI and a novel metric D is contained within the spread between historical runs, and the response to the changing forcing is well described by the trends. After 2100, however, three of the five models considered undergo a significant weakening in the strength of all major areas and links (and therefore overall connectivity) in the only ensemble member available, while IPSL and MPI-EMS show an increase in the overall strength of areas, both in the tropics and at high latitudes, pointing to the large uncertainty in the predictability of the long term evolution of climate modes.

Bibliography

[1] Knutti R. & Sedláček J. (2013) Robustness and uncertainties in the new CMIP5 climate model projections *Nature Clim. Change* **3**, 369–373

[2] Fyfe J. C., Gillett N.P. & Zwiers F. W. (2013) Overstimated global warming over the past 20 years *Nature Clim. Change* **3**, 767–769

[3] Fountalis, I. Bracco, A. & Dovrolis, C. (2015) ENSO in CMIP5 simulations: network connectivity from the recent past to the twenty-third century *Clim. Dyn.* On-Line Early Release, DOI 10.1007/s00382-014-2412-1

[4] Cheng, W., Chiang, J. C.H. & Zhang, D. (2013) Atlantic Meridional Overturning Circulation (AMOC) in CMIP5 Models: RCP and Historical Simulations. *J. Climate* **26**, 7187–7197.

[5] Chadwick, R., Boutle, I., & Martin, G. (2013) Spatial Patterns of Precipitation Change in CMIP5: Why the Rich Do Not Get Richer in the Tropics. *J. Climate* **26**, 3803–3822.

[6] Tett, S.F.B., M.J., Cartis, C., Rowlands, D.J. & Liu, P. (2013) Can top of atmosphere radiation measurements constrain climate predictions? part 1: Tuning. *J. Climate*, **26**, 9348–9366

[7] Sherwood, S.C., Bony, S. & Dufrense, J.-L. (2014) Spread in model climate sensitivity traced to atmospheric convective mixing. *Nature*, **505**, 37–42

[8] Barnes, E. A. & Polvani, L. (2013) Response of the midlatitude jets, and of their variability, to increased greenhouse gases in the CMIP5 models *J. Climate*, **26** 7117–7135

[9] Little, C. M., Horton, R. M., Kopp, R. E. (2015) Uncertainty in twenty-first-century CMIP5 sea level projections. *J. Climate*, **28**, 838–852

[10] Turner, J., Bracegirdle, T. J., Phillips, T., Marshall, G. J. & Hosking, J. S. (2013) An initial assessment of Antarctic Sea Ice extent in the CMIP5 models. *J. Climate*, **26**, 1473–1484

[11] Kay, J. E., Hillman, B. R., Klein, S. A., Zhang, Y., Medeiros, B., Pincus, R., Gettelman, A., Eaton, B., Boyle, J., Marchand, R. & Ackerman, T. P. (2012) Exposing global cloud biases in the Community Atmosphere Model (CAM) using satellite observations and their corresponding instrument simulators. *J. Climate*, **25**, 5190–5207.

[12] Kharin, V. V., Zwiers, F. W., Zhang, X. & Wehner M. (2013) Changes in temperature and precipitation extremes in the CMIP5 ensemble *Climate Change*, **119**, 345–357

[13] Stevens, B. & Bony, S. (2013) What are climate models missing? *Science* **340**, 1053–1054

[14] Fountalis, I. Bracco, A. & Dovrolis, C. (2014) Spatio-temporal network analysis for studying climate patterns. *Clim. Dynam.* **42**, 879–899

[15] Neelin, J. D., Bracco, A., Luo, H., McWilliams, J. C., & Meyerson, J. E. (2010) Considerations for parameter optimization and sensitivity in climate models. *Proc. Natl. Acad. Sci. USA*, **107**, doi:10.1073/pnas.1015473107

[16] Bracco, A., Neelin, J. D., Luo, H., McWilliams, J. C., & Meyerson, J. E. (2013) High dimensional decision dilemmas in climate models *Geosci. Model Dev.* , **6**, 1673–1687

[17] Neale, R.B. & co-authors (2013) Description of the NCAR Community Atmosphere Model (CAM 5.0) NCAR Tech. Note TN-486+STR, 268 pp., Natl. Cent. for Atmos. Res., Boulder, Colo., available at http://www.cesm.ucar.edu/models/cesm1.0/cam/

[18] Park, S. & Bretherton, C.S. (2009) The University of Washington shallow convection and moist turbulence schemes and their impact on climate simulations with the Community Atmosphere Model. *J. Climate*, **22**, 3449–3469

[19] Gettelman, A., Morrison, H. & Ghan S.J. (2008) A new two-moment bulk stratiform cloud microphysics scheme in the Community Atmospheric Model (CAM3), Part II: single-column and global results *J. Clim.*, **21**, 3660–3679

[20] Park, S., Bretherton, C. S. & Rasch, P. J. (2014) Integrating Cloud Processes in the Community Atmosphere Model, Version 5. *J. Climate*, **27**, 6821–6856

[21] Bretherton, C.S. & Park, S. (2009) A new moist turbulence parameterization in the Community Atmosphere Model. *J. Climate*, **22**, 3422–3448

[22] Iacono, M.J., Delamere, J.S., Mlawer, E.J., Shephard, M.W., Clough, S.A. & Collins, W.D. (2008) Radiative forcing by long-lived greenhouse gases: Calculations with the AER radiative transfer models *J. Geophys. Res.*, **113**, D13103

[23] Bogenschutz, P.A., Gettelman, A., Morrison, H., Larson V.E., Schanen, D.P., Meyer, N.R. & Craig, C. (2012) Unied parameterization of the planetary boundary layer and shallow convection with a higher-order turbulence closure in the Community Atmosphere Model: single-column experiments *Geosci. Model Dev.*, **5**, 1407–1423

[24] Hourdin, F., Grandpeix, J.-Y., Rio, C., Bony, S., Jam, A., Cheruy, F., Rochetin, N., Fairhead, L., Idelkadi, A., Musat, I., Dufresne, J.-L., Lahellec, A., Lefebvre, M.-P. & Roehrig, R. (2013) LMDZ5B: the atmospheric component of the IPSL climate model with revisited parameterizations for clouds and convection emphClim. Dyn., 40, 2193–2222

[25] Yang, B., Qian, Y., Lin, G., Leung, L.R., Rasch, P.J., Zhang, G. J., McFarlane, S.A., Zhao, C., Zhang, Wang, Y.H., Wang, M. & X. Liu (2013) Uncertainty quantification and parameter tuning in the CAM5 Zhang-McFarlane convection scheme and impact of improved convection on the global circulation and climate *J. Geophys. Res. Atmos.*, **118**, 395–415

[26] Severijns, C.A. & Hazeleger, W. (2005) Optimizing Parameters in an Atmospheric General Circulation Model. *J. Climate*, **18**, 3527–3535

[27] Murphy, J.M., Booth, B.B., Collins, M., Harris, G.R., Sexton, D.M., Webb, M.J. (2007) A methodology for probabilistic predictions of regional climate change from perturbed physics ensembles. *Philos Trans A Math Phys Eng Sci.*, **365**, 1993–2028

[28] Knight C.G., Knight, S.H.E., Massey, N., Aina, T., Christensen, C., Frame, D.J., Kettleborough, J.A., Martin, A., Pascoe, S., Sanderson, B., Stainforth, D.A. & Allen, M.R. (2007) Association of parameter, software, and hardware variation with large-scale behavior across 57,000 climate models *Proc. Natl. Acad. Sci. USA*, 104, 12259–12264

[29] Rougier J., Sexton D.M.H., Murphy J.M., Stainforth D. (2009) Analyzing the climate sensitivity of the HadSM3 climate model using ensembles from different but related experiments. *J. Climate*, **22**, 3540–3557

[30] Rowlands, D. J. & co-authors (2012) Broad range of 2050 warming from an observationally constrained large climate model ensemble *Nature Geoscience*, **5**, 256–260

[31] Jackson, C.S., Sen, M.K., Huerta, G., Deng, Y. & Bowman, K. P. (2008) Error reduction and convergence in climate prediction *J. Climate*, **21**, 6698–6709

[32] Prieß, M., Koziel, S. & Slawig, T. (2011) Surrogate-based optimization of climate model parameters using response correction *J. Comput. Science*, **2**, 335–344

[33] Hairer, M. & Majda, A.J. (2010) A simple framework to justify linear response theory. *Nonlinearity*, **23**, 909–922

[34] Bellprat, O., Kotlarski, S., Luthi, D. & Schär, C. (2012) Objective calibration of regional climate models *J. Geophys. Res.-Atmos.*, **117**, D23115

[35] Archibald, R., Chakoumakos, M., & Zhuang, T. (2012) Characterizing the elements of Earths radiative budget: applying uncertainty quantification to the CESM *E*lsevier Science Journal, Procedia Comput. Sci., **9**, 1014–1020

[36] Kalnay, E., Kanamitsu, M., Kistler, R., Collins, W., Deaven, D., Gandin, L., Iredell, M., Saha, S., White, G., Woollen, J., Zhu, Y., Chelliah, M., Ebisuzaki, W., Higgins, W., Janowiak, J., Mo, C., Ropelewski, C., Wang, J., Leetmaa, A., Reynolds, R., Jenne, R. & Joseph, D. (1996) The NCEP/NCAR 40-Year Reanalysis Project *Bull. Amer. Meteorol. Soc.*, **77**, 437–471

[37] Bracco A., Kucharski, F., Kallummal, R. & Molteni F. (2004) Internal variability, external forcing and climate trends in multi-decadal AGCM ensembles *Climate Dynamics*, **23**, 659–678

[38] Tsonis, A. & Roebber, P. (2004) The architecture of the climate network. *Physica A*, **333**, 497–504

[39] Newman, M., Barabasi, A.L. & Watts, D.J. (2006) *The structure and dynamics of networks*. Princeton University Press, 592 pp.

[40] Tsonis, A.A., Swanson, K. & Kravtsov, S. (2007) A new dynamical mechanism for major climate shifts. *Geophys. Res. Lett.*, **34**, L13705

[41] Tsonis, A.A. & Swanson, K. (2008) Topology and predictability of El Nio and La Niña networks. *Phys. Rev. Lett.*, 100, 228502

[42] Yamasaki, K.A., Gozolchiani, A. & Havlin S. (2008) Climate networks around the globe are significantly affected by El Niño. *Phys. Rev. Lett.*, 100, 228501

[43] Gozolchiani, A., Yamasaki, K., Gazit, O. & Havlin, S. (2008) Pattern of climate network blinking links follows El Niño events. *Europhys. Lett.*, **8**3, 28005

[44] Donges, J.F., Zou, Y., Marwan, N. & Kurths, J. (2009) The backbone of the climate network. *Europhys. Lett.*, **8**7, 48007

[45] Steinhaeuser, K. & Tsonis, A.A. (2014) A climate model intercomparison at the dynamics level. *Clim. Dynamics*, **42**, 1665–1670

[46] Kawale, J., Liess, S., Kumar, A., Steinbach, M., Snyder, P., Kumar, V. & Semazzi, F. (2013) A graphbased approach to find teleconnections in climate data. *SADM*, **6**, 158–179

[47] Donges, J.F., Schultz, H.C., Marwan, N., Zou, Y. & Kurths, J. (2011) Investigating the topology of interacting networks. *Eur. Physics J. B*, **8**4, 635–651

[48] Donges, J.F., Zou, Y., Marwan, N. & Kurths, J. (2009) Complex networks in climate dynamics. *Eur. Physics J. -Special Topics*, **174**, 157–179

[49] Yamasaki, K.A., Gozolchiani, A. & Havlin, S. (2009) Climate networks based on phase synchronization analysis track El-Niño. *Prog. Theor. Phys. Supp.*, **1**79, 178–188

[50] Holland, P.W. (1986) Statistics and causal inference. *J. Am. Statist. Assoc.*, **8**1, 945–960

[51] Steinhaeuser, K., Chawla, N.V. & Ganguly, A.R. (2011) Complex networks as a unified framework for descriptive analysis and predictive modeling in climate science. *SADM*, **4**, 497–511

[52] Rayner, N. & co-authors (2003) Global analyses of sea surface temperature, sea ice, and night marine air temperature since the late nineteenth century. *J. Geophys. Res-Atmos. 1984–2012* 108, D14, 27

[53] Gent, P. & co-authors (2011) The Community Climate System Model version 4. *J. Climate*, **2**4, 4973–4991

[54] Hubert, L. & Arabie, P. (1985) Comparing partitions. *J. Classif.*, **2**, 193–218

[55] Steinhaeuser, K. & Chawla, N.V. (2010) Identifying and evaluating community structure in complex networks. *Pattern Recog. Lett.*, **31**, 413–421

[56] Taylor, K.E., Stouffer, R.J. & Meehl, G.A. (2012) An Overview of CMIP5 and the Experiment Design. *Bull. Amer. Meteorol.Soc.*, **93**, 485–498

[57] Martin, G.M. & co-authors (2011) The HadGEM2 family of Met Office Unified Model climate configurations *Geosci. Model Dev.*, **4**, 723–757

[58] Miller, R.L. & co-authors (2014) CMIP5 historical simulations (1850-2012) with GISS ModelE2. *J. Adv. Model. Earth Syst.*, **6**, 441–477

[59] Smith, T.M., Reynolds, R.W., Peterson, T.C. & Lawrimore. J. (2008) Improvements to NOAAs historical merged landocean surface temperature analysis (18802006). *J. Climate*, **21**, 2283–2296

[60] Carton, J.A. & Giese, B.S. (2008) A reanalysis of ocean climate using simple ocean data assimilation (SODA). *Mon. Weather Rev.*, **136**, 2999–3017

[61] Meinshausen, M., Smith, S.J., Calvin, K., Daniel, J.S., Kainuma, M.L.T., Lamarque, J.F. & van Vuuren, D.P.P. (2011) The RCP greenhouse gas concentrations and their extensions from 1765 to 2300. *Climate Change*, **109**, 213–241

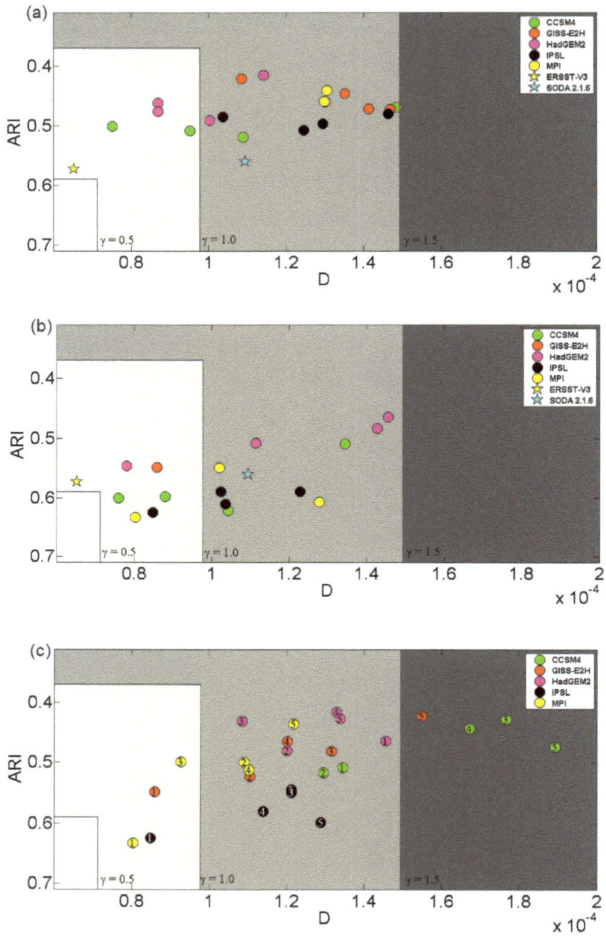

Figure 4. Metric D versus ARI for networks constructed using JJA sea surface temperature fields in 5 climate models participating to CMIP5 (a) during the period 19562005, for up to 4 ensemble members for each model, (b) during 2051-2100 for the same members, and (c) from 2051 to 2300 over five consecutive 50-year periods, from 1 to 5, for the only ensemble member extending past 2100. In the historical period networks are referenced to the HadISST and the metrics are also indicated for two other reanalysis products. In the projected simulation all networks are referenced to the corresponding integration over the historical period. In the middle panels the metrics for the reanalysis products are repeated for context. Three levels of noise-to-signal ratios γ are also indicated.

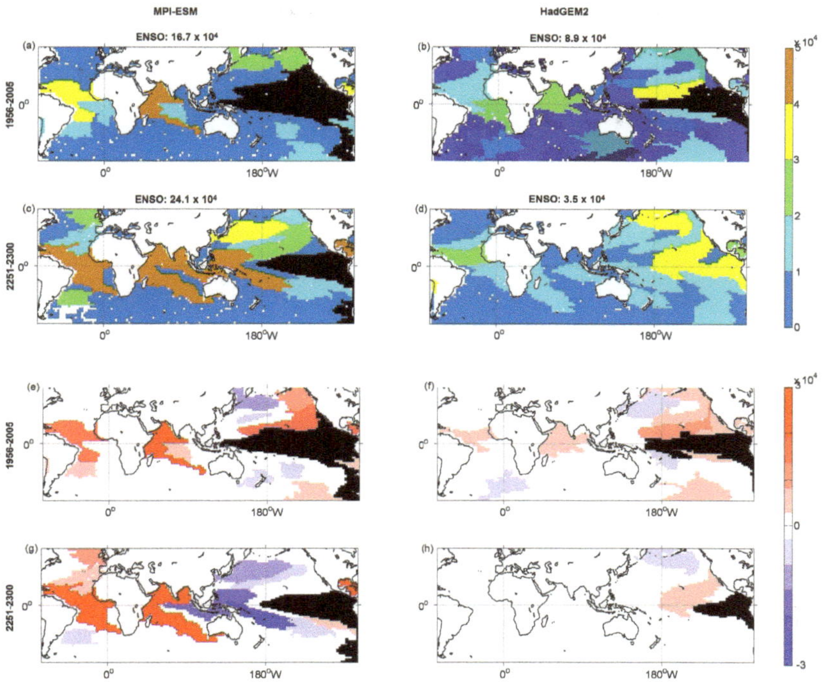

Figure 5. Sea surface temperatures strength maps (a-d) and link maps from the ENSO area (e-h) in JJA for two models, MPI-ESM (left) and HadGEM2 (right) over the historical period (1956-2005) and in the distant future (2251-2300).

Climate dynamics on global scale: resilience, hysteresis and attribution of change

Klaus Fraedrich [†], Isabella Bordi [‡] and Xiuhua Zhu [*]
[†] Max Planck Institute of Meteorology, Hamburg, Germany
[‡] University of Rome, La Sapienza, Rome, Italy
[*] University of Hamburg, Hamburg, Germany

Abstract The dynamics of a set of zero- and one-dimensional Energy Balance Models (EBM) of the Earth's climate are subjected to a systematic analysis of the response to changes of the greenhouse effect in terms of atmospheric opacity (effective emissivity), which is related to CO_2 concentration. (*i*) Experiments with abrupt CO_2-decrease (opacity increase) from the actual value are performed with the one-dimensional EBM, which reveal the following results: There is a critical opacity threshold beyond which the model ends up at a modern snowball Earth. It occurs within a few percentage changes around this threshold because the model strongly depends on the relationship among atmospheric temperature and the sudden ice-albedo feedback activation. Two different time scales characterize the temperature decline; the first almost linear part appears to be related to the change of the opacity and the corresponding sharp increase of the planetary albedo; the second part is mainly controlled by the rapid change of the planetary albedo towards its higher value (ice-albedo feedback). Return from snowball Earth requires a large CO_2 concentration as only the ice-albedo feedback is effective while the water vapor feedback is ineffective at low temperatures. Introducing the ice margin as an EBM-variable, hemispheric asymmetry can be identified for different polar albedos while keeping the meridional heat transports. (*ii*) Experiments with transient opacity changes reveal three kinds of hysteresis cycles when introducing cyclic changes: static, dynamic and memory hysteresis. Hysteresis loops are attained for a cyclic change of the greenhouse forcing between the extreme conditions (near bifurcation points) of maximum (almost all outgoing infrared radiation is trapped by the atmosphere) and minimum greenhouse effect (when almost all the infrared radiation is emitted to space). Static hysteresis loops are attained for vanishing heat capacity while dynamic hysteresis shows delay due to the ocean's heat capacity. For cyclic change far from bifurcation points, the model climate depends only

A. Provenzale et al. (Eds.), *The Fluid Dynamics of Climate*,
CISM International Centre for Mechanical Sciences
DOI 10.1007/ 978-3-7091-1893-1_6 © CISM Udine 2016

on the history of the radiative forcing thus displaying a hysteresis cycle that is neither static nor dynamical (it does not include bifurcation points nor shows a cyclic steady state), but is related to the memory response of the model determined by the heat capacity (ocean mixed-layer depth).

1 Introduction

Common methods of climate model analysis are sensitivity experiments to determine the response to small variations of the external forcing, supposed the system is in a steady state. But such analyses would be misleading when the system has a few steady states. In fact, a small change in the forcing (for example around a CO_2 threshold) could lead to a dramatic change in the steady state so that the analysis would not be able to capture the complexity of the system's responses. Therefore, in this contribution to the fluid dynamics of climate we like to extend the traditional sensitivity studies of the climate system analyzing the problem of no return to a fixed forcing and provide a systematic evaluation of two types of forcing change, abrupt and cyclic.

The aim of this chapter[1] is to present – in a comprehensive way – results and novel interpretations of climate dynamics on global scale, that is on resilience, hysteresis and attribution of change; we are focusing on the present day snowball Earth tipping point as it is obtained by changes of greenhouse gas forcing (as described by Bordi et al., 2012, 2013; Fraedrich, 2012). First, the global climate system is introduced as an energy balance model (EBM, Section 2). Here it should be noted that, when long time scales are analyzed, its chaotic nature is averaged out (Held et al., 2010) and only residuals emerge. In this case, a simple model of the surface energy budget can account for most of the responses. Secondly, two characteristic response experiments are conducted and evaluated, both of which affect the greenhouse feedback of the climate system (Section 3); that is, the responses to abrupt and to transient-cyclic changes of the greenhouse forcing. A systematic analysis of the dynamical system reveals three classes of hysteresis: static, dynamic and memory hysteresis. The loop of the latter can be interpreted as a novel phenomenon associated with a dynamic interpretation of resilience, when the system is not affected by tipping points.

[1] A summary of our recent research

2 The global climate in a box: Energy Balance Model

To demonstrate climate modeling strategies, a minimum model is introduced first to provide settings for climate analyses. It is based on a poor man's radiation scheme leading to the greenhouse climate system which, when in equilibrium, plays a similar role in climate dynamics as geostrophy and hydrostasy do in geophysical fluid flow. First, the dominant radiative fluxes contributing to the climate are defined in terms of a simple two-stream method, which is reduced to an atmosphere interacting with the land/ocean surface by radiative fluxes only; that is, sensible and latent heat fluxes are not explicitly considered. Parameterization of the atmosphere by statistically deduced feedbacks leads to the statistic-dynamical model version, which is extended to include feedbacks, transient hystereses, and meridional exchanges (and it can be subjected to stability, sensitivity, and stochastic analyses) in order to characterize climate variability.

2.1 Dynamical core

Coupling fast atmospheric dynamics with a slow land/ocean, requires a special modeling strategy to explicitly resolve the dynamics of the slow system. A common approach is to parameterize the influence of the fast compartment, which leads to the *statistic-dynamical* climate model with feedbacks incorporating the statistical effects of the fast system. A similar strategy is employed for Global Circulation Models (GCMs) when parameterizing fast and small-scale processes of the boundary layer or cloud ensembles (after suitable space/time-averaging).

Radiative scheme: A simple two-stream radiation scheme is introduced with an atmospheric (subscript A) and land/ocean or surface (no subscript) layer associated with the respective heat capacities C_A and C. Long-wave or terrestrial radiative fluxes are described by the Stefan-Boltzmann law σT^4 with $\sigma = 5.67\ 10^{-8}\ \mathrm{Wm^{-2}K^{-4}}$. The long-wave upward flux from a black body surface, σT^4, is absorbed in the upper layer (absorption coefficient α), while the long-wave downward radiation is totally absorbed by the surface. The incoming short-wave or solar radiation, $S_0 = 340\ \mathrm{Wm^{-2}}$, passes a completely transparent atmosphere; it is absorbed at the surface from where a remaining part is reflected to space, $R \downarrow = S_0(1 - a)$, with the (planetary) albedo or whiteness a. Absorption of solar radiation in the atmosphere would require an additive term contributing to the radiative heating. This leads to a minimum energy balance model of a greenhouse climate with a black body land/ocean surface and without solar absorption in the atmosphere:

$$\textit{Atmosphere} \qquad C_A(T_A)_t = 0 + \alpha\sigma T^4 - 2\alpha\sigma {T_A}^4$$

$$\textit{Land/Ocean} \qquad CT_t = S_0(1-a) - \sigma T^4 + \alpha\sigma {T_A}^4$$

The subscript t attached to the temperatures of the atmosphere and the surface, T_A and T, denotes the time derivative. At equilibrium (C, $C_A = 0$) the incoming solar radiation balances the long-wave outgoing radiation. First, the atmosphere is considered leaving the land/ocean fixed to provide the boundary conditions ($C = 0$). Its dynamics is relatively fast due to the small heat-capacity, $C_A = c_p(\Delta p/g)$, compared to the ocean heat capacity, $C = 7$ Wm^{-2} K^{-1} year, estimated for a depth of 50 m. The incoming solar radiation S_0, planetary albedo a, and emissivity α, lead to a stable equilibrium with $R_o = R \downarrow = R \uparrow$. The global surface temperature $T = T_A \sqrt[4]{2} \simeq$ 288 K satisfies $S_0(1-a) = \sigma T^4(1 - \frac{1}{2}\alpha)$, and exceeds the atmosphere's by about 20%. This characterizes the greenhouse effect and corresponds to a vertical temperature difference between surface and atmosphere of $T - T_A \simeq$ 40 K. Note that atmospheric temperature, $T_A \simeq 250$ K, to which all initial conditions converge, corresponds to the observed temperature vertically averaged over the atmospheric mass $2\Delta p$, ${T_A}^* = (2\Delta p)^{-1} \int T_A dp$, which is close to the observed mid-troposphere temperature near 500 hPa. Linear stability, $(\delta T_A)_t = -\delta T_A/\tau_A$, of the equilibrium solution is determined by Newtonian cooling with the radiative time scale, $\tau_A = \frac{1}{4} C_A T_A/[\alpha S_0(1-a)]$ of 1 to 2 months.

Dynamical core: A suitable Ansatz for the parameterization is a diagnostic atmosphere, $C_A = 0$, which reacts "instantaneously" on changes of the slow system and feeds back to it by the Stefan-Boltzmann effect. This defines the dynamical core of the statistic-dynamical energy balance model (EBM) with effective emissivity (opacity) $\epsilon = (1-b)$ and $b = \frac{1}{2}\alpha$.

$$\textit{Dynamical core} \qquad CT_t = S_0(1-a) - \epsilon\sigma T^4 \qquad (1)$$

The knowledge of T at climate steady state ($C = 0$) will allow estimating the atmospheric opacity ϵ by knowing S_0 and a. For the present climate $a = 0.3$ and $T = 288$ K so that $\epsilon = 0.61$. For a snowball Earth, instead, we assume $a = 0.7$ and $T = 211$ K which provides a steady solution for $\epsilon =$ 0.9. The difference in infrared emission is about 130 Wm^{-2}. By instantaneously decreasing CO_2 to 20 ppm, we increase the long-wave emission of the atmosphere by just 15 Wm^{-2}. Thus, to reach a snowball Earth requires other greenhouse gases, namely the water vapor and its temperature related feedback (see Bordi et al., 2012, and Section 2.2).

2.2 Feedbacks and parameterizations

Outgoing and incoming radiation $R\uparrow = \epsilon\sigma T^4$ and $R\downarrow = S_0(1-a)$ include feedback processes affecting the atmospheric emissivity (infrared opacity) ϵ and surface albedo a. They need to be included in the dynamical core describing the statistical (long-term) effects of fast variables on slower ones. Regression-type linear feedbacks are most commonly used in climate modeling modifying the greenhouse climate by its temperature dependent albedo and opacity:

Ice-albedo effect: This positive feedback links surface albedo ($0 < a < 1$) and surface temperature.
Temperature-albedo: *temperature drop → more snow → higher albedo*
Albedo-temperature: *less SW-radiation absorbed → further temp. drop*

Various formulations have been employed: The simplest formulations are (i) a quadratic form $a = a_2 - b_2T^2$, (ii) a hyperbolic tangent function of T ranging from $a = 0.3$ to 0.7, $a = 0.5 - 0.2\tanh\frac{(-20-T)}{10}$, and (iii) a step function interval for the temperature (with T in degree Celsius) with $a = 0.3$ for $T > -10°$C and $a = 0.7$ for $T < -10°$C, which may also describe the limits of the hyperbolic tangent function.

Greenhouse effect: This positive feedback associates effective emissivity ($0 < \epsilon < 1$) with the surface temperature, and thus with the atmospheric moisture content as a prominent greenhouse gas.
Temperature-moisture: *temp. rise → more evaporation → more vapor*
Moisture-temperature: *more IR radiation from sky → further temp. rise*

Various formulations are being employed: (i) A positive greenhouse-water vapor feedback, $\epsilon = 1 - b = \epsilon(T)$, which is related to the surface temperature, has been suggested from clear sky radiation measurements in the tropics, $b = c_2 + d_2T^2$. (ii) Considering CO_2-emittance only, similar parameterizations are being used based on logarithmic relationships. For example $c_2 = 0.0235\ln(CO_2)$ with CO_2 concentration in parts per million by volume (ppm); or assigning a reference value of $C_0 = 280$ ppm, $c_2 = 0.1 + 0.007\ln(CO_2/C_0)$ (Singer et al., 2014); and the CO_2-forced heating relation for GCMs, $5.35\ln(CO_2/C_0)$ with a reference $C_0 = 360$ ppm, which has been discussed in IPCC. That is, by decreasing CO_2 to 20 ppm (keeping the temperature fixed), the long-wave emission is decreased by merely 15 Wm^{-2}.

Classical applications – resilience, tipping points and attribution: Employing the quadratic parameterizations of both ice-albedo and greenhouse feedbacks, the EBM introduced here reveals a gradient system (Fig. 1) which has originally been used to demonstrate resilience and catastrophes or tipping points in a zero-dimensional climate system (see Fraedrich, 1979, and Fig. 1): Resilience due to changes in the state variables and in external parameters (forcing) can be measured by suitable integration of the gradient system (or its potential) along the T-trajectory from repellor to climate mean at fixed forcing (Fig. 1a) or in parameter space from climate mean to relevant tipping point (Fig. 1b), respectively. Recently this model has been linked with a set of regional socio-economic subsystems to provide a global change analysis incorporating econometric measures of change attribution and to understand the complexity of domestic and international interactions under the global change perspective (see Singer et al., 2014), after adapting the model's performance to past climates and future climate scenarios.

2.3 From zero to one dimension

First, we simplify the outgoing long-wave flux parameterization by the first two terms of the binomial expansion of T^4 around the reference temperature of 273 K with T in degree Celsius

$$CT_t = S_0(1-a) - \epsilon(A+BT) \tag{2}$$

The coefficients $A = 315$ Wm^{-2}, $B = 4.6$ Wm^{-2} °C^{-1} are estimated for an effective emissivity (opacity) $\epsilon = 1$. Anomalies about the reference temperature, $(\delta T)_t = -{\tau_S}^{-1}\delta T$, show its internal stability to which all initial conditions converge. The (negative) eigenvalue represents the slow relaxation time scale $\tau_S = \frac{1}{4}CT/[S_0(1-a)] \simeq 10$ to 20 months $\gg \tau_A$. To include ice margins and meridional transports latitudinal dependence needs to be incorporated; for example, the position of the ice line, cannot be accounted for by the 0-dimensional EBM. However, a surface energy budget that takes this feature into account needs to include meridional heat fluxes $F(T)$ by a simple parameterization (Lindzen, 1990):

$$CT_t = S_0(1-a) - \epsilon(A+BT) + F(T) \tag{3}$$

with $F(T) = k(T^* - T)$, where T^* and k are the global mean surface temperature and the heat-flux coefficient, respectively. For S_0 we use the annual average latitudinal dependence (see Bordi et al., 2012, loc. cit.):

$$S_0 = \frac{1}{4}1367(1 - \frac{1}{2}0.477(3x^2-1))$$

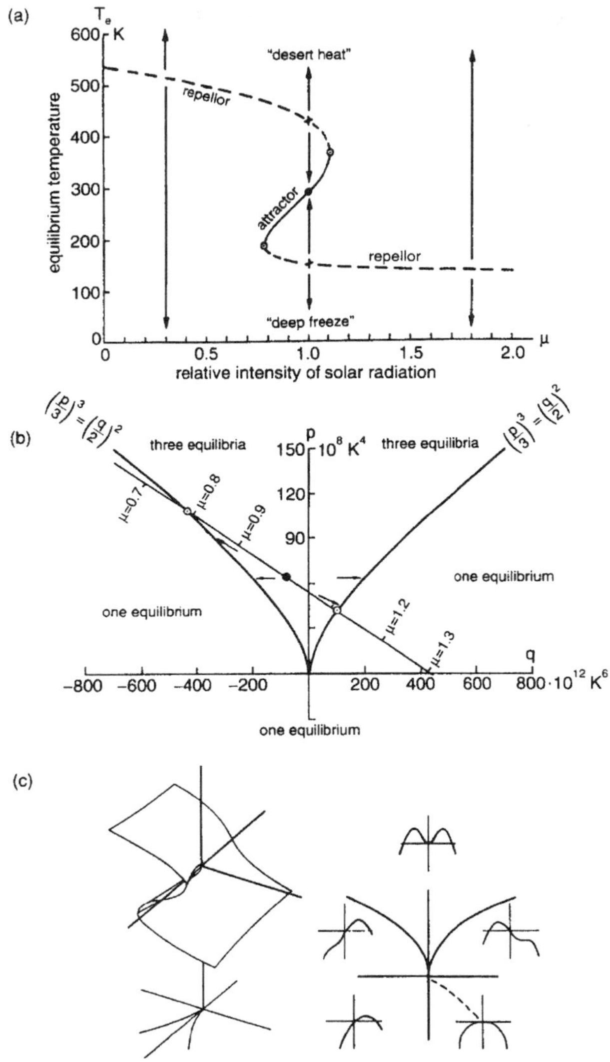

Figure 1. Equilibria of a zero-dimensional EBM climate model (gradient system) with ice albedo and greenhouse feedback: (a) state space spanned by the equilibrium surface temperature versus the changing relative solar radiation, (b) bifurcation diagram spanned by generalized (p,q)-axes comprising all external parameters, and (c) schematic presentation of the potential of a gradient system in the generalized parameter space (see Fraedrich, 1979).

where $x = \sin(\phi)$ with ϕ being the latitude. Now, with the ice margin temperature $T(X) = -10°C$ at latitude X, the exchange coefficient $k = 2.2$ Wm^{-2} °C^{-1}, and the opacity $\epsilon = 0.61$ (for present climate conditions) we obtain $X = 0.95$. To obtain the global mean surface temperature the planetary albedo at the ice margin, $a(X) = (a_1+a_2)/2$, is required satisfying (in a simple format) the ice-albedo feedback by the step function Ansatz with $a_1 = 0.3$ and $a_2 = 0.62$ below and above the ice margin, respectively. The value of a_2 has been chosen to be the averaged surface ice albedo in the Northern Hemisphere (NH) for a snowball Earth state (Bordi et al., 2012, taken from PlaSim simulations). This leads to the global mean surface temperature:

$$T^* = \int_0^1 (S_0(1 - a(x))/\epsilon B - A/B)dx$$

Summarizing, this EBM box model setting provides a toy-modeling approach for dynamical systems diagnostics of abrupt and cyclic changes, expanding the classical tipping point and stochastic analyses.

3 Dynamics of hysteresis and resilience: abrupt and cyclic changes

The dynamical model core (Eq. 1) with its suite of parameterizations (Eqs. 2 and 3) will be employed to uncover details of tipping point dynamics which have not been discussed in the classical analyses (in the seventies) and the more recent rediscovery about thirty years later. As tipping point we select the transition to modern snowball Earth by changing the greenhouse feedback induced by the terrestrial radiation contribution to the Earth's energy balance. In contrast to the commonly analyzed effect of changing solar radiation input (due to the Milankovich cycles) we change the CO_2-related terrestrial radiation affecting the dynamics induced by ice-albedo and water-vapor greenhouse feedbacks. *Abrupt changes* are analyzed first (Bordi et al., 2012): (i) What are the time scales before reaching what appears to be a bifurcation or tipping point towards modern snowball Earth? (ii) How functions the back-transition from the modern snowball Earth, when the CO2 concentration is set to very large values keeping the solar constant at present value? (iii) How are radiative perturbations related to the latitude of the ice margin? (iv) And are there hemispheric asymmetries induced by radiative effects, that is cooling one hemisphere more than the other? *Cyclic changes* are analyzed next (Bordi et al., 2013) to account for hysteresis effects, which range from (i) static via (ii) dynamic to (iii) memory hysteresis.

3.1 Abrupt change dynamics

The setting to analyze the dynamics induced by abrupt changes is the dynamical core of the 0-dimensional EBM (Eq. 1) amended by the ice-albedo and the greenhouse feedback parameterizations and the extension to a 1-dimensional system introduced in the previous section on model development.

Towards snowball Earth: The dynamical core (Fig. 2a, Eq. 1) combined with the hyperbolic-tangent ice-albedo feedback function of T (ranging from $a = 0.3$ to 0.7) is integrated with increasing the initial value from present day $\epsilon = 0.61$ up to 0.9 at a given time rate to mimic the long wave radiative effect of the CO_2. This corresponds to a reduction towards 20 ppm where the transition to snowball Earth occurs. The transition from low albedo values to the higher one at 0.7 occurs at the threshold $T = -20°$C. The time-dependent solution shows that the surface temperature (after adjusting to perturbation) decreases (i) almost linearly from its initial value of about 15°C toward about −10°C and then (ii) when it crosses the threshold of the major albedo change at −20°C (at $a = 0.7$), it quickly decreases to about −60°C. This is the equilibrium temperature if all parameters are kept constant from then on. That is, two different time scales characterize the temperature decline: (i) The first part appears to be related to the change of the opacity ϵ and the corresponding sharp increase of the planetary albedo. (ii) The second part is mainly controlled by the rapid change of the planetary albedo toward its higher value (ice-albedo feedback). Note that this simple EBM is able to capture the same features of the surface temperature behavior as provided by the GCM PlaSim (see Bordi et al., 2013).

Return from snowball Earth: The dynamical core reveals a range of external parameters ϵ (or S_0) characterized by the co-existence of two stable solutions: snowball and temperate Earth. This range depends on (i) the threshold temperature at which the planetary albedo changes and (ii) the sharpness of this transition (Fig. 2b). This is shown by the temperature steady states when varying ϵ back and forth within the opacity ϵ-interval 0.2 - 0.9: The planetary albedo step function ($a = 0.3$ and 0.7, black line) shows that the transition between these boundary values occurs at $T = -10°$C. The planetary albedo hyperbolic tangent function (blue and red lines) centered at $T = -10°$C and $T = -20°$C, respectively. This shows that if the system is in a snowball state, and if we increase a greenhouse gas such as CO_2, we might escape from this state provided that the associated radiative forcing ϵ reaches the threshold value identified by the lower branch

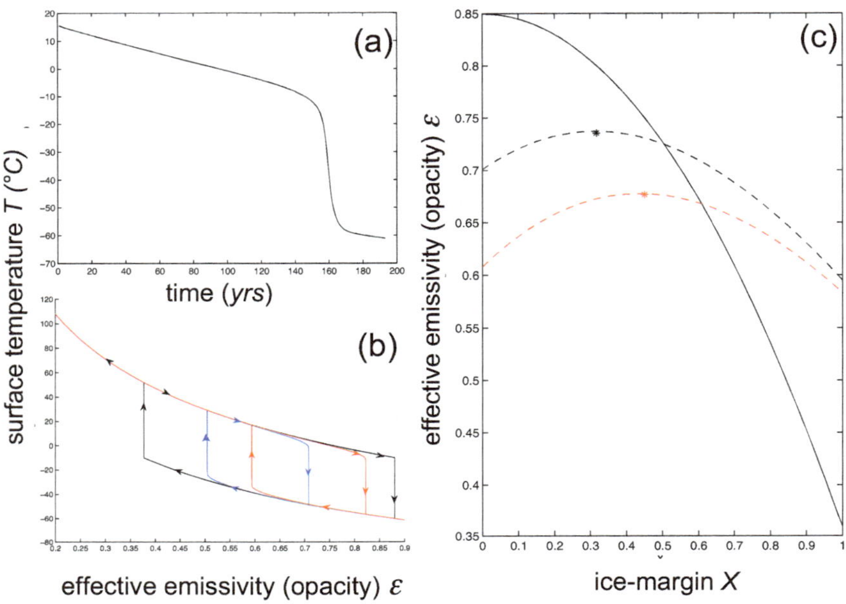

Figure 2. EBM analysis: (a) Time evolution of the surface temperature T obtained by integrating the 0-dimensional energy balance model (Eq. 2) for the opacity ϵ varying linearly from 0.61 to 0.9. The parameters settings are listed in the text. (b) Steady solutions (Eq. 2) for the surface temperature T as a function of ϵ varying back and forth in the interval 0.2-0.9. The shape of the planetary albedo is specified in the text. Arrows denote the direction of variation of ϵ. Units are dimensionless and degree Celsius for ϵ and T, respectively. (c) Radiative forcing ϵ as a function of the ice margin X (solution of Eq. 4) for $\beta = 0$ and $a_2 = 0.62$ (solid black line), $\beta > 0$ and $a_2 = 0.62$ (dashed black line), $\beta > 0$ and $a_2 = 0.7$ (solid red line). Asterisks denote the maximum of the curves for $\beta > 0$. Units are dimensionless (see Bordi et al., 2013).

of the hysteresis curves. Moreover, it is predicted that the new state will be a great deal warmer than what is expected because we must add also the water vapor effect that dominates once the evaporation has been re-established and it involves the insulating nature of ice. Also, the rate of approach to the two steady state branches is just a consequence of the sharpness of the albedo change associated with the ice-albedo feedback activation or deactivation. In particular, Fig. 2b (red line) shows the transition from snowball to ice-free Earth to occur for $\epsilon = 0.59$ in case of setting the T-bound to $-20°C$. This value of the greenhouse forcing must be provided by adding only CO_2, since the water vapor effect is excluded due to lack of evaporation. This explains why the large value of CO_2 concentration is needed for the transition back from the ice-covered state.

Radiation – ice margin relation: The latitudinal dependence of the ice line can only be considered by a 1-dimensional EBM (Fig. 2c, Eq. 3). Normally, we expect that the ice margin moves towards the equator ($X = 0$) as the radiative forcing ϵ increases and vice-versa. The presence of heat transports should destabilize this unique relationship (ice cap instability) by introducing a critical ice margin X^* so that for any $X < X^*$ there is only the snowball solution. Thus, to study the behavior of ϵ as a function of X, all the other parameters are fixed. It turns out that by evaluating Eq. 3 at X (with $C = 0$ or steady state condition), the opacity ϵ must satisfy the following diagnostic equation:

$$\epsilon^2\Lambda_1 + \epsilon^2\Lambda_2 + \Lambda_3 = 0 \qquad (4)$$

where the coefficients Λ_i ($i = 1, 2, 3$) are functions of the parameters evaluated at a given X. In Fig. 2c (black lines) the solutions as a function of X are shown in case of $\beta = 0$ (solid line) and $\beta \neq 0$ (dashed line). Note that for a given X there are two real solutions; the negative one must be excluded since $\epsilon > 0$. Note also that in case of no heat transports, for each ϵ-value in the interval 0.35-0.85 within the bifurcation points (roughly the same range is identified in Fig. 2, black line) there is a unique value of X. When heat transports are active, they cool the equatorial regions and warm the polar ones. This implies that lower values of ϵ are sufficient to allow the ice to advance towards the tropics, while the opposite holds at high latitudes. Moreover, it appears that a critical X exists for which the solution is in a snowball state (the ice forms also in the tropics). In our case, $X^* = 0.32$ (or 18.7°N) which corresponds to the maximum values of ϵ. This is the critical latitude beyond which the ice is expand to the equator. Moreover, it appears that for ϵ between 0.7 and 0.74 (maximum value) two ice margins satisfy the diagnostic equation. Since the heat fluxes introduce a critical ice margin so that the ice can form also in the tropics, they are a key parameter for the transition to snowball state and vice versa, and a characteristic feature of the dynamics of this snowball Earth tipping point.

Hemispheric asymmetry: Since in the SH the surface albedo of the ice is expected to be higher than in the NH (as supported by GCM simulations when the ice-covered state is reached), let us consider the solution of the diagnostic equation with the same heat transports but $a_2 = 0.7$ (red dashed line in Fig. 2c). In this case, the range of variability of the radiative forcing associated with the ice margin is reduced and, most importantly, the critical ice margin is changed to $X^c = 0.45$, which corresponds to about 26.7°N. This means that the snowball Earth solution is reached more rapidly for a higher value of a_2 because the critical latitude of the ice margin is moved poleward. This explains asymmetries between NH and SH, which is particularly evident when ozone is removed. In concluding, the origin of the asymmetric response between the NH and SH lies in the different ice albedo of the two hemispheres.

In summarizing, sudden decrease of atmospheric trace gases that interact with thermal radiation shows the following effects: If there are bifurcation points, the dynamics depends crucially on the exact amount of the gas concentrations as long as their effect on water vapor is significant. There appears to exist a CO_2 threshold leading to a modern snowball Earth. The response of the thermal field has a great degree of simplicity; that is, when long time scales are analyzed with the chaotic behavior being averaged out only residuals occur (Held et al., 2010). Thus, a simple energy balance model may account for many aspects of climate dynamics.

3.2 Cyclic change dynamics – hysteresis and resilience

The response of a zero-dimensional energy balance model (Eq. 2, linearized dynamical core) is studied. It includes the ice-albedo feedback by the step-function approach and the greenhouse effect expressed by effective surface emissivity, $\epsilon = (1 - b)$; the outgoing infrared radiation changes between the extreme conditions of maximum and minimum greenhouse effect, $b \in [1 > b_1 > b_2]$, representing cyclic changes of CO_2-concentration, which contrasts the commonly studied Milankovic cycles affecting the incoming solar radiation cycles. The following cycle is restricted not to include the bifurcation points:

$$\beta(t) = \beta_{01} - bt \quad \text{for } t \text{ from } (n-1)\tau \text{ and } n\tau \text{ for } n = 1, 3, 5, ...$$
$$\beta(t) = \beta_{02} + b(t - \tau) \quad \text{for } t \text{ from } (m-1)\tau \text{ and } m\tau \text{ for } m = 2, 4, 6, ...$$

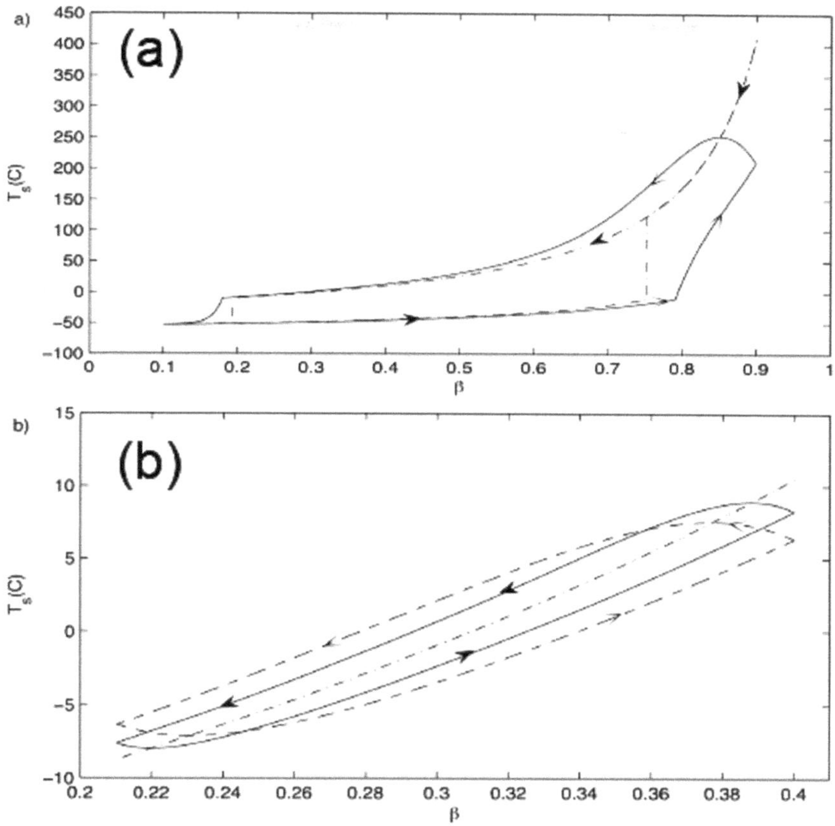

Figure 3. EBM solutions in (β, T)-plane: (a) Static hysteresis loop (dash-dotted) for $C = 0$ and β varying statically back and forth in the interval [0.9, 0.1], and dynamical hysteresis loop (solid line) for $C = 50$ Wm^{-2} K^{-1} year and β varying in the same interval at the rate $b = 0.001$ year^{-1}. (b) Locus of steady states (dash-dotted) for $C = 0$ and β varying statically in the interval [0.4, 0.21], and memory hysteresis loops for $C = 50$ Wm^{-2}K^{-1}year (solid), $C = 100$ Wm^{-2} K^{-1}year (dashed) and β varying in the same interval at the rate $b = 0.001$ year^{-1}. Units are dimensionless and in degree Celsius for β and T (see Bordi et al., 2013).

with the half-cycle time $\tau = (\beta_1 - \beta_2)/b$, and the boundary conditions $\beta_1 = \beta((n-1)\tau) = 0.9$ and $\beta_2 = \beta((m-1)\tau) = 0.1$ determine β_{01} and β_{02}. The selected boundary values represent the extreme conditions of maximum greenhouse effect (almost all outgoing infrared radiation is trapped by the atmosphere) and of minimum greenhouse effect (when almost all the infrared radiation is emitted to space). Two "cyclic" EBM experiments are performed, that is one with and one without including the bifurcation points into the cycle.

Cycle with bifurcation points: So far, two types of hysteresis experiments are being distinguished.
Static hysteresis: If the heat capacity C is set to zero and β varies extremely slowly (adiabatically) back and forth within in the selected β-interval, then a static hysteresis loop appears (Fig. 3a, dash-dotted). This corresponds to the conventional time slice approach performed by global climate model experiments analyzing climate change under varying radiative forcing conditions.
Dynamic hysteresis: For $C = 50$ Wm^{2}K^{-1}yr and β varying back and forth in the same β-interval at the rate $b = 0.001$ yr^1, a dynamical hysteresis loop emerges (Fig. 3a, solid curve). Within the β-range the values of radiative forcing, which lead to the transition from ice-free to ice-covered Earth state and vice versa, can be easily estimated from the upper and lower branches of the hysteresis cycles. Note that by construction, such transitions occur when there is a change in the albedo a, i.e. the surface temperature crosses the threshold of $-10°$C. Note that the dynamical hysteresis shows different transition points compared to the static hysteresis.

Cycle without bifurcation points: There is a dynamical hysteresis without bifurcation. To illustrate this case, consider β within [0.4, 0.21] so that the system is linear with $T > -10°$C and $a = 0.35$. Note that $\beta =$ 0.4 is an estimation of the greenhouse parameter for present climate ($a =$ 0.35 and $T = 15°$C); the value of $\beta = 0.21$, instead, has been estimated from Fig. 3a as lower bound of β before transition to the ice-covered Earth state. For $C = 0$, the solution of the EBM will follow the continuous set of steady states determined by the β-value at a given time without static hysteresis. For $C \neq 0$, instead, the solutions will depart from the steady states because of the delayed response of the system induced by the heat content (l.h.s. of EBM). This departure is expected to be positive when β decreases and negative when β increases. Since the end points are fixed, the two branches will join there, so that the full cycle follows a closed curve in the (β, T)-plane.

Memory hysteresis and resilience: Define β as input, the linear core as transducer, and T as observation, then the solution is a dynamical hysteresis loop but without a static counterpart. Therefore, such a loop cannot be attributed as static or dynamical hysteresis (as in Goldsztein et al., 1997), because it does not represent the locus of the steady states and it does not include the bifurcation points. Therefore, we coin it memory hysteresis which is due to a kind of memory effect. That is, the system responds on a longer time scale compared to the forcing change by a time delay of adjustment to the time-dependent forcing variation. Given a time rate of change of radiative forcing, the source of this memory effect lies in the heat capacity C whose magnitude controls the width of the hysteresis loop. For this case entropy is not a well-defined quantity, as the area enclosed in the loop will not equal the entropy production. Instead, the area is a measure of the resilience (and of the memory) to the time-dependent change of the forcing. The EBM trajectories are displayed as hysteresis loops in a (β, T)-space for a given time rate b and two values of heat capacity C (Fig. 3b). The dash-dotted curve is the locus of the model steady states ($C = 0$), while the time-dependent solutions for decreasing (increasing) β are shown for large (dashed) and small (full) heat capacities. Three features are noted: First, for different greenhouse gas parameters we obtain the same temperature. Second, when the model switches from decreasing to increasing β (or vice versa) the model response lags the forcing so that the surface temperature will continue to increase (or decrease) despite the fact that the forcing is already decreasing (or increasing). Third, an increasing heat capacity C enhances the effect of the system's memory broadening the memory hysteresis loop and thus enhancing resilience.

Recalcitrant response: The lagged memory response is not associated with the recalcitrant or slow response (see Held et al., 2010): Two heat capacities characterize the dynamics, a shallow ocean layer responding rapidly to the atmosphere, and a deeper ocean which are responsible for the fast and slow components observed in the GCM response, though it has a similar effect. The slow component of the global warming refers to a climate model response towards equilibrium by an instantaneous return to the preindustrial radiative forcing. The problem of no "return" to a fixed forcing which our systematic hysteresis analyses are related to, is different, as the forcing is cyclically changed at a given time rate and transient (not steady) solutions are followed. Moreover, this EBM has a single time scale, which is determined by the heat capacity C (consistent with GCM-PlaSim simulations). For any rate of CO_2 change, the relaxation towards equilibrium for $CO_2 = 20$ ppm is characterized by a single time scale dictated by the depth of the slab ocean. Thus, the slow component has a different origin in the two cases: In the study by Held et al. (2010) it is related to the slow adjustment of the system towards equilibrium to a sudden reduction of the radiative forcing, while in the present study it is due to the delay of the system in adjusting to the time-dependent forcing variations. The area enclosed by the loop may be a relevant parameter to be considered when greenhouse gases are projected into different future values; climate, in fact, may be trapped by loops like these.

4 Conclusions

A climate system with bifurcation points as been introduced has an energy balance model associated with ice-albedo and greenhouse-temperature feedbacks. Its dynamics has been analyzed by a set of experiments to characterize the response to abrupt and transient-cyclic greenhouse forcings. The novel features revealed in an EBM environment have also been observed in global climate model simulations following the same experimental design (as described by Bordi et al., 2012, 2013; Fraedrich, 2012). This substantiates the observation that, when long time scales are analyzed, the chaotic nature of the climate system is averaged out (Held et al., 2010) and only residuals emerge. The results presented in this lecture note are in parts a linear combination of a set papers (Bordi et al., 2012, 2013; Fraedrich, 2012). The aim is twofold: First, to extend climate system analyses to transient forcing and present the methods and results in the energy balance model framework. Secondly, to note that the design of comprehensive global climate model experiments, in particular when employing transient forcings, should be preceded by a thorough analysis of a simpler surrogate climate system of the type of energy balance models, which are subjected to the same suite of conditions.

Acknowledgements

We thank Professor Alfonso Sutera for decade-long collaboration, continuing discussions, inspirations and exchanges.

Bibliography

Bordi, I., K. Fraedrich, A. Sutera, X. Zhu, *Transient response to well-mixed greenhouse gas changes*, Theor. Appl. Climatol., **108**, 245-252, doi:10.1007/s00704-011-0580-z, 2012.

Bordi, I., K. Fraedrich, A. Sutera, X. Zhu, *On the effect of decreasing CO_2 concentrations in the atmosphere*, Clim. Dyn., **40**, 651-662, doi:10.1007/s00382-012-1581-z, 2013.

Fraedrich, K., *Catastrophes and resilience of a zero-dimensional climate system with ice-albedo and greenhouse feedback*, Q. J. Roy. Meteorol. Soc., **105**, 147-167, 1979.

Fraedrich, K., *A suite of user-friendly global climate models: Hysteresis experiments*, Eur. Phys. J. Plus, **127**, 127:53, doi: 10.1140/epjp/i2012-12053-7, 2012.

Goldsztein, G. H., F. Broner, S. H. Strogatz, *Dynamical hysteresis without static hysteresis: scaling laws and asymptotic expansions*, SIAM J. Appl. Math., **57**, 1163-1187, 1997.

Held, I. M., M. Winton, K. Takahashi, T. Delworth, F. Zeng, G. K. Vallis, *Probing the fast and slow components of global warming Transient response to well-mixed greenhouse gas changes by returning abruptly to preindustrial forcing*, J. Climate, **23**, 2418-2427, doi:10.1175/2009JCLI3466.1, 2010.

Lindzen, R. S., *Dynamics in Atmospheric Physics*, Cambridge, University Press, New York, 1990.

Singer, C., T. Milligan, T. S. Gopi Rethinaraj, *How China's options will determine global warming*, Challenges, **5**, 1-25, doi:10.3390/challe5010001, 2014.

Water in the climate system

Elisa Palazzi [†] and Antonello Provenzale [‡]
[†] Institute of Atmospheric Sciences and Climate, CNR, Torino, Italy
[‡] Institute of Geosciences and Earth Resources, CNR, Pisa, Italy

Abstract Water is an essential element of climate, owing to its dynamical and thermodynamical effects. Water is, in addition, one of the main fluids active in the climate system: oceans, ice sheets and glaciers, snow and atmospheric water (in vapor, liquid or solid form) and the whole hydrological cycle are at the heart of the workings of the Earth System. In these short notes, we shall describe just a few of the many ways water enters the fluid dynamics of climate.

1 The water cycle

One of the special characteristics of our Planet is that the physical conditions at or near its surface are not too far from the triple point of water, which means that water can be easily found in solid, liquid or vapor form. Phase transitions of water are common, with the associated absorption and release of latent heat and the corresponding moist thermodynamical processes. In turn, (liquid) water, as a polar solvent, allows for complex biogeochemical reactions and it is one of the requisites for the presence of life, at least in the form we know.

Water thus plays a crucial role in the Earth's climate, and it goes through a complex set of transformations and passages (including phase transitions) known as the "water cycle". The vast reservoirs of liquid water in the oceans undergo evaporation, thus providing water vapor in the atmosphere. The vapor is transported vertically (through convection) and horizontally (through advection) in the atmosphere, and it condensates into cloud droplets and ice crystals (releasing latent heat). These can, in turn, aggregate into raindrops, snowflakes or hail, that in due course precipitate on the Earth surface because of the gravity pull. Snow feeds high mountain glaciers and polar ice sheets, and it covers wide areas of our planet during the winter months. Liquid water, from rain or snow/glacier melt, flows on continental surfaces (runoff) or into the ground, forming water reservoirs and aquifers which can reach considerable depth. Most of this water eventually flows back to the ocean, while some of it goes deeper down in the Earth's crust, encountering

A. Provenzale et al. (Eds.), *The Fluid Dynamics of Climate*,
CISM International Centre for Mechanical Sciences
DOI 10.1007/ 978-3-7091-1893-1_7

higher temperatures and possibly being re-ejected at the surface through cracks in the rocks and geothermal activities.

Some of the water in continental areas is locally evaporated, and much more is transpired by vegetation, that represents an important source of moisture to the atmosphere. In continental areas far from the oceans, the contribution of evapotranspiration from soils and plants is a major player in the local hydrological cycle, even though, overall, on our Planet the role of oceans as moisture sources is unequalled. For the sake of exploring alternative realities, some authors have imagined a planet covered with soil, with water in it, but without oceans. The question was, then, whether such a hypothetical planet (reminiscent of Arrakis in the Dune science fiction book series) could host vegetation; that is, whether plant roots could extract enough water from the soil to sustain a hydrological cycle. The answer, at least for a very simplified model, was positive (Cresto Aleina et al., 2013).

Fluid dynamics and thermodynamics are at the core of the workings of the water cycle. In all its states, water is characterized by movement: liquid water flows in ocean currents, river and lake circulations, water motion takes place in porous media (aquifers). In vapor form, water is transported by the atmospheric currents and it contributes to the atmospheric dynamics (especially convection) thanks to its phase transitions. In solid form, ice flows - albeit slowly - from the center of the big ice sheets to their edges, or from the top of mountain glaciers to their snout.

Changes in climate are reflected in complex modifications of the water cycle, and, correspondingly, changes in some aspects of the water cycle can have profound effects on climate. The water cycle of interglacial periods (such as the one we are living in now) is very different from the water cycle of glacial epochs. During the last glacial period, vast expanses of ice covered large portions of Europe and North America, while the lower latitudes were much dryer than today. The distribution of precipitation was rather different, and the water cycle was characterized by rapid variations (Margari et al., 2010).

In more recent epochs, changes in precipitation (amount, seasonality and distribution) triggered important changes and had a profound effect on human societies (Beniston, 2010; Trenberth, 2011; IPCC, 2014). Such changes were caused by many different instances, such as changes in the distribution of Sea Surface Temperature, changes in the monsoons, and maybe modifications in the land cover and vegetation.

Today, global temperature rise, although relatively modest when compared to past changes, is nevertheless quite rapid and it is accompanied by changes in precipitation patterns (amount, intensity and spatial/temporal distribution) and in the storage of water in the different reservoirs: mountain glaciers and Arctic sea and continental ice are rapidly disappearing (but Antarctic sea ice is growing), and the snow cover is decreasing, both in extent, depth and duration (see the latest IPCC report for a full account of the ongoing changes; IPCC, 2013). Part of the continental melted ice is now in the ocean, and this, together with the increase in the ocean water volume owing to the increased temperature (thermal expansion), leads to sea level rise.

Precipitation, in particular, is a crucial variable for human societies: too much of it, and we have to face floods and extreme events. Too little of it, and we are exposed to droughts and possibly famine. For this reason, understanding how the water cycle will change in different global change scenarios is a crucial requirement. In some sense, we can say that precipitation is the most important climatic variable for human well-being. Unfortunately, however, precipitation (and in general all variables related to the water cycle) are the most difficult to measure and model. Uncertainties are still large, even in characterizing current conditions, and a large amount of work is required to provide a better understanding of the Earth's water cycle and its ongoing changes.

In this notes, we shall tackle only some of the issues related to the water cycle, focusing on those which are the closest to our personal research activities, and on the uncertainties which characterize water variables. Many more interesting and challenging things could be said, but these will remain, at least here, unspoken about.

2 Changes in precipitation

Precipitation is a crucial driver of the water cycle. On continental regions, precipitation feeds glaciers and ice sheets, surface runoff in rivers and streams and subsurface aquifers, and it is essential for vegetation growth. In addition, precipitation is one of the climate variables which has the most direct impact on human activities, including agriculture and drinking water.

In the last decades, precipitation displayed changes in its spatial distribution. Figure 1, from Turco et al. (2015), shows the observed differences in average annual precipitation amount, ΔP, between the period 1981-2010 and the period 1951-1980, for winter (left panel) and summer (right panel), indicated as a percentage change with respect to the 1951-1980 values. Precipitation is estimated on a global grid with resolution 5x5 degrees using the GPCC dataset version 6 (Schneider et al., 2014). Similar results are obtained using other gridded datasets.

Contrary to temperature rise, which is observed globally albeit with important regional differences (think of the Arctic), precipitation displays a complex pattern of regional changes, with areas that become rainier and others that become drier. In addition, the patterns are often different in winter and summer. The first lesson which we can learn from observations in the last 60 years is thus that precipitation changes display huge spatial and temporal variability.

A very important signal emerges in sub-Saharan Africa, where one observes a significant decrease in precipitation. Here, long-standing drought conditions, combined with geopolitical difficulties, have generated critical situations for the people living in the area.

For other areas, the situation is more complex, with precipitation changes having opposite signs in different seasons (for example, Australia). In the Mediterranean, there is a tendency towards dryer conditions, especially for southern Europe, as already indicated by several previous analyses (e.g. Hoerling et al., 2012; Spinoni et al., 2015, and references therein).

Globally, it is not easy to say what is happening to the hydrological cycle and to its intensity. Higher atmospheric temperatures imply (as described by the Clausius-Clapeyron equation) higher saturation water pressure, that is, a potentially larger amount of water vapor in the air. Since water vapor is a powerful greenhouse gas, its increase would lead to still larger temperatures activating a strong positive feedback. However, the larger saturation pressure only allows for a larger amount of moisture, it does not imply it. In order to get more water in the atmosphere, larger evaporation from the ocean (or evapotranspiration from land) is required. One could well imagine a case in which the total amount of water vapor remains constant, while relative humidity (the ratio between the effective moisture content of air and the content at saturation) decreases.

In fact, available observations suggest instead that relative humidity remained approximately constant, thus, that the total amount of water vapor in the atmosphere increased following the growth of the saturation water pressure and of air temperature (Held and Soden, 2000; Sherwood and Meyer, 2006; McCarthy et al., 2009). Consequently, one expects a positive feedback between temperature and atmospheric moisture, as well as an intensification of the water cycle: more moisture means more energy and more water for precipitation, associated with stronger evaporation.

But did the hydrological cycle really intensified? Many authors addressed this point. Giorgi et al. (2011) defined an index measuring hydroclimatic intensity, as HY-INT $=$ INT $\times$ DSL where INT is the average rainfall intensity (that is, annual rainfall divided by the number of rainy days) and DSL is the mean annual dry spell length, that is, the mean number of consecutive days in a dry spell. The analysis of several model outputs indicates that climate projections foresee an increase of HY-INT for most

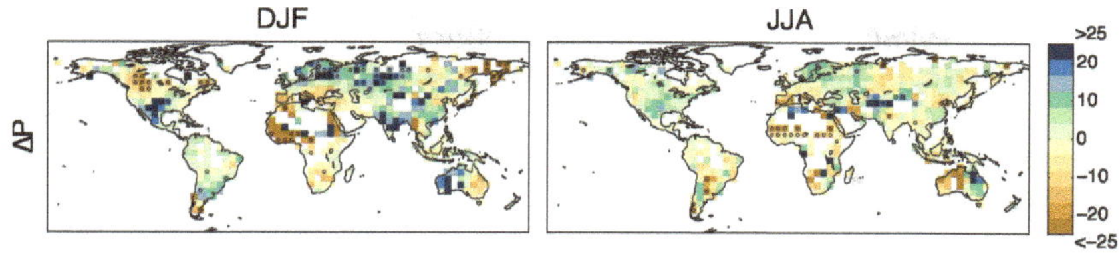

Figure 1. Percentage changes in December-January-February (DJF, left panel) and June-July-August (JJA, right panel) in the mean precipitation (P). The changes are calculated between the periods 1981- 2010 and 1951-1980 from the GPCC dataset. Black circles indicate significant (at 95% level) changes. From Turco et al. (2015).

regions of the world. However, the analysis of observed data in the last few decades provides mixed results: while some regions such as Europe have witnessed an increase of HY-INT, the southern hemisphere generally did not show an increase in the intensity of the hydrological cycle. In general, this index displays rather different trends in different regions and it is not easy to determine whether there is a significant tendency at global level.

Besides averages, societal end environmental challenges are even more sensitive to extremes, in terms of the properties of the high tail of precipitation intensity, the structure of the spatial pattern of rainfall extremes and the maximum duration of dry spells and/or heat waves. Future projections obtained from climate models tend to indicate an increase in these parameters, but current observational data do not provide an univocal picture, especially because extreme events are rare (luckily). In addition, high-resolution, regional models provide precipitation statistics (and thus distributions of extremes) that depend quite heavily on the convective and microphysical parameterizations adopted (Pieri et al., 2015).

Once more, these little notes illustrate how difficult is to identify trends and patterns of change associated with the hydrological cycle. Caution is needed, especially given the large uncertainties in our knowledge of water cycle variables (e.g., precipitation), as illustrated in the following section.

3 Water cycle in the Hindu-Kush Karakoram Himalaya and the role of uncertainty

Precipitation is one of the most important but difficult variables of the water cycle, both to model and measure, especially in areas with complex orography such as mountain regions. Mountains act as natural reservoirs of water, "storing" precipitation from the cold season, when most of it falls as snow and forms snowpack, until the warm season when snowpack melts feeding streams and rivers and releasing water for a number of different uses. Snowpack also adds mass to glaciers in their accumulation zone. Precipitation is thus a central glacier-making ingredient. Therefore, a good representation of the precipitation amounts, spatial distribution and characteristics in high-altitude regions is important for a better understanding of the cryosphere system as well as for assessing the impacts of climate variability and change on water resources, hydrological risks, mountain ecosystems and biodiversity and finally on downstream societies relying on high-altitude water resources.

The Hindu-Kush Karakoram Himalaya (HKKH) is the largest mountain area in the world, feeding some of the major rivers in Southeast Asia which bring water to more than 1.5 billion people. The position, characteristics and variety of the orographic reliefs in the area represent major climatic drivers. The region is characterized by the interaction of local and large-scale circulation systems that give rise to a precipitation climatology, and by consequence glacier dynamics and hydrological regimes, that are strongly dependent on the geographic region, as well as on the season. In particular, two main (climatically homogeneous) sub-regions can be distinguished in the HKKH area (Palazzi et al., 2013): the Himalayan range in the eastern portion of the region is exposed to the summer (June to September) monsoon precipitation, while the westernmost part, the Hindu-Kush Karakoram (HKK), is exposed to the arrival of westerly midlatitude perturbations bringing precipitation during winter and early spring as well as, but to a smaller extent, to the summer monsoon circulation (Archer and Fowler, 2004; Syed et al., 2006; Yadav et al., 2012). These different seasonal water inputs give rise to a different climatological annual cycle of precipitation in the two sub-regions: in the HKK the precipitation distribution is bimodal and reflects the wintertime precipitation associated with the westerly perturbations and the impact of the summer monsoon, while in the Himalaya the precipitation distribution is unimodal with a peak around July. Overall, this picture is coherently reproduced by various datasets based on in-situ station and satellite data, or a combination of both, by reanalyses and in some cases also by global climate model (GCM) data (see e.g. Palazzi et al.,

2013, 2015). However many uncertainties remain in both measuring and modelling precipitation in the mountains and in particular in the HKKH region. Precipitation estimates obtained from sensors based on different measurement techniques can differ widely from each other and can be affected by large biases, leading to significant uncertainties in precipitation estimates and also to model evaluations and validations that can be biased by the dataset used as a reference.

Figure 2 from Palazzi et al. (2013) shows, as an example, the spatial distribution of summer (June to September, JJAS) precipitation over a region that includes the HKKH range obtained from three datasets based on the interpolation of in-situ station data (APHRODITE, CRU, GPCC), one satellite dataset (TRMM), a merged raingauge and satellite climatology (GPCP), the ERA-Interim reanalyses and data from one state-of-the-art GCM (EC-Earth; Hazeleger et al., 2012). Precipitation is averaged over the period 1998-2007 for which data from all datasets are available. Although the key features of the summer precipitation field over the target area are well represented by all datasets (maximum precipitation amounts over the eastern stretch of the Himalaya and decreasing values from southeast to northwest along the Himalayan chain), important discrepancies arise from the different temporal and spatial sampling and resolution and from the specific characteristics of the various products (e.g., different bias correction, homogenization or interpolation choices). It is also important to point out that while the reanalysis and GCM data estimate the total precipitation (including snow), the APHRODITE, CRU, and GPCC station data and the TRMM satellite product provide rainfall estimates because both rain gauges and satellite sensors do not measure (or strongly underestimate) the solid component of precipitation (Rasmussen et al., 2012). In areas with sparse station coverage, moreover, gridded datasets such as GPCC, APHRODITE, and CRU, interpolate grid-point values from the nearest few available stations, which constitutes one of the major limitations and potential sources of uncertainty in this kind of data.

On the other hand, also simulating seasonal precipitation and its spatial and temporal variability in mountain regions has proven to be a difficult task. One reason is the need to parameterize the large-scale effects of the processes that occur below the scales that the models explicitly resolve (Lal et al., 2000): radiation, heat transfer, cloud microphysics, the planetary boundary layer, and deep convection are among the processes commonly included in global (and, in most cases, even regional) climate models by means of parameterizations. Additionally, other aspects of the model formulation make it difficult to correctly simulate precipitation: as an important example, the coarse vertical and horizontal resolution does

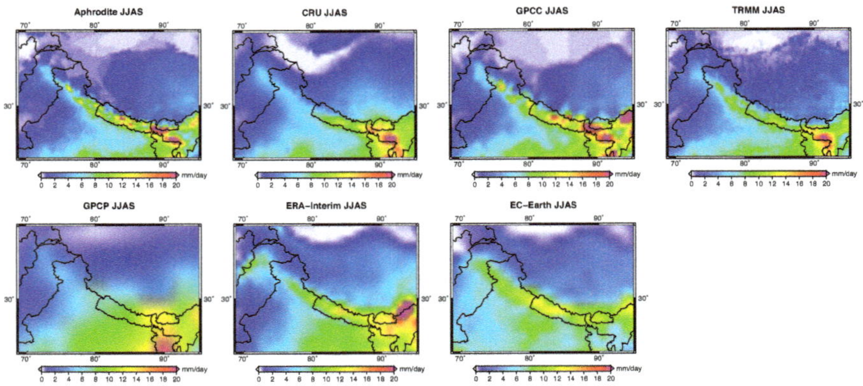

Figure 2. Multiannual mean (1998-2007) of summer (June to September, JJAS) precipitation over the region between 69°E-95°E and 23°N-39°N from the APHRODITE, CRU, GPCC, TRMM, GPCP, ERA-Interim, and EC-Earth model datasets. From Palazzi et al. (2013).

not allow to adequately represent regional forcings associated with steep topography (such as orographic lifting). On short time scales, precipitation is also influenced by aerosol particles that indirectly affect cloud properties (Kaufman and Fraser, 1997), a particularly severe problem in the monsoon-dominated areas like the Himalayas. In a recent study, Palazzi et al. (2015) assessed the ability of thirty-two state-of-the-art GCMs participating in the latest Coupled Model Intercomparison Project (CMIP5) to reproduce the current and future precipitation climatology in the HKKH region. This study shows that the multi-model mean (MMM) of the CMIP5 ensemble and most individual models exhibit a wet bias with respect to the observational datasets taken as reference (CRU and GPCC) in both the HKK and Himalaya regions and for all seasons. Such positive bias is commonly found in the precipitation simulated by state-of-the-art GCMs over high-elevation regions; the Tibetan Plateau, in particular, constitutes one of the most outstanding examples of such model behaviour (Su et al., 2012). The reasons for the general positive precipitation model bias could be attributed either to the model inaccuracies in representing small-scale processes, mainly owing to the model spatial resolution and imperfect parameterizations (Lee et al., 2010; Su et al., 2012), or to shortcomings of the station and gridded data used for comparison, which tend to underestimate the solid component of precipitation (Rasmussen et al., 2012; Palazzi et al., 2013), or to a combination of the two aspects. Palazzi et al. (2015) also found differences in the GCMs spread depending on the season and the considered sub-region (HKK or Himalaya), with the models differing greatly in the climatology of precipitation which they produce in the HKK region in summer. These authors performed a cluster analysis by grouping the models providing a similar representation of the precipitation annual cycle in the two regions by using a hierarchical clustering technique, in order to identify the group or groups of models performing better compared to the observations and their common features (if any). The analysis produced four models clusters, whose characteristics are shown in Fig. 3. In the Himalaya (top panels), all clusters produce unimodal precipitation annual cycles with different widths and amplitudes. The models belonging to Cluster 1 reproduce, on average, a precipitation annual cycle very close to the observed one, especially in summer. In the HKK (bottom panels) the four model clusters produce precipitation annual cycles with very different characteristics and only one group of models (Cluster 4) produces an annual cycle similar to the observed one (bias aside), whose peaks in late winter/early spring and summer reflect the two main seasonal precipitation sources in the area. However, no feature of the better-performing models in the Himalaya or in the HKK region has definitely emerged as one playing a pivotal role for providing the best results in terms of precipitation annual cycle in the two regions (Palazzi et al., 2015).

Since no single model can be chosen as "the best one", it is certainly important to use results from the whole range of models (IPCC, 2013). However, given the large differences in the basic model outputs, a multimodel ensemble estimate should be regarded with extreme caution (Tebaldi and Knutti, 2007).

4 Long-distance moisture transport and local evaporation: dynamics of the Western Weather Patterns

As mentioned in the previous section, winter/spring precipitation over the HKK mountain range, in the western Himalayas, is affected by westerly perturbations, also known as western weather patterns (WWP). In the western Himalaya and in the Karakoram, WWPs are primarily responsible for the build-up of seasonal snow cover, which represents a crucial water reservoir and a significant source for some of the major river basins in the region (Archer and Fowler, 2004); the study by Bookhagen and Burbank (2010)

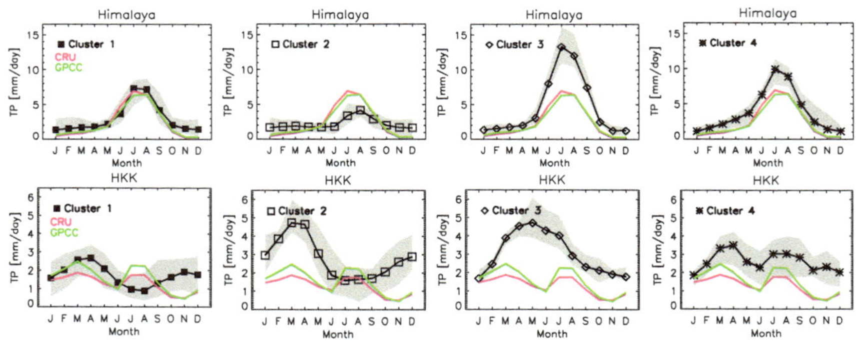

Figure 3. Mean annual cycle of precipitation in the Himalaya (top panels) and in the HKK (bottom panels) region simulated by all models within each of the four identified clusters (the grey shaded areas indicate the variability range of the models) and by their MMM. The CRU and GPCC observations are shown as the pink and green lines, respectively. Adapted from Palazzi et al. (2015)

shows that snowmelt constitutes up to 50% of the total annual discharge in this area.

Recent studies have shown that the North Atlantic Oscillation (NAO), the dominant pattern of atmospheric variability in the North Atlantic sector affecting climate across much of the Northern Hemisphere and across Europe, is an important regulating factor also in the Karakoram region (Syed et al., 2006; Yadav et al., 2009). In fact, these studies have found that winter precipitation in the HKK and the NAO are correlated, with larger precipitation amounts typically recorded during the positive NAO phase. It has been suggested that the NAO regulates winter precipitation in this region by strengthening the Middle East jet stream (MEJS) from North Africa to southeastern Asia during its positive phase. The stronger jet intensifies the WWPs at longitudes between 40°E and 70°E, where the majority of moisture transport toward the HKK takes place (Yadav et al., 2009), so that faster westerlies in the middle to lower troposphere intensify moisture transport to the Karakoram (Syed et al., 2010). Evaporation plays an important role in this mechanism (Filippi et al., 2014): during the positive NAO phase, enhanced evaporation occurs from the northern Arabian Sea, the Red Sea, and the Persian Gulf - mainly related to higher surface wind speed - and, to a lesser extent, from the Mediterranean Sea. The increased humidity due to evaporation combines with the intensification of westerlies to give enhanced moisture transport toward the Karakoram.

Changes in the spatial structure of the NAO pattern can be crucial in determining the strength of the correlation between the NAO and precipitation in the HKK and to understand the multidecadal variations that this correlation underwent during the twentieth century (Yadav et al., 2009). By using the Angle Index (AI) defined by Wang et al. (2012) to measure the slow movements of the NAO centers of action (COAs), Filippi et al. (2014) showed that significant positive correlations between the NAO and precipitation in the HKK occur for low values of this Angle Index (negative or at least very small tilt of the NAO) because in this situation the NAO exerts a strong control on the MEJS and, as a consequence, the mechanism of regulation of the HKK precipitation by the NAO is activated. Even though sources of variability other than the NAO for precipitation in the HKK have been identified that can add noise to the NAO-precipitation signal in the HKK in winter (e.g. Yadav et al., 2010), it is clear that the position of the NAO COAs has regulated the strength of the NAO-precipitation relationship in the HKK region during the last century.

5 Soil-vegetation-atmosphere water fluxes

On Planet Earth, atmospheric moisture is largely generated by evaporation from the oceans. On continental areas, however, evaporation from the soil and, most importantly, transpiration from plants (Jasechko et al., 2013) can provide a significant contribution to the water content of the atmosphere and they represent an important component of the water cycle.

On a given portion of land, atmospheric water comes from two sources: lateral transport (column-averaged convergence of the horizontal atmospheric moisture fluxes) and evapotranspiration (evaporation plus transpiration) from the lower surface (the soil-vegetation system). Atmospheric moisture is lost from that portion of the atmosphere by lateral transport (column-averaged divergence of the horizontal moisture flux) and condensation and aggregation processes leading to precipitation reaching the soil (precipitation that re-evaporates before reaching the ground affects the vertical distribution of moisture but not its total amount in the column). Notice that, at this level of description, we do not distinguish between the different phases of water provided they do not fall to the ground. From this balance, for a vertical column over a certain surface area S we can write

$$\frac{dW}{dt} = -\frac{1}{S} \int_{\sigma} \rho_w \, \vec{u} \, d\sigma + \mathrm{ET} - \mathrm{P} \qquad (1)$$

where W is the atmospheric water content (in kg/m^2), $\rho_w \vec{u}$ is the horizontal moisture flux across the surface of the column σ (including the lateral and top surfaces but excluding the bottom boundary), with ρ_w the density of water (in vapor, liquid or solid phase) and $\vec{u}$ is the wind velocity. Both ρ_w and $\vec{u}$ are functions of space and time, and the usual convention is that the flux is positive when it is carrying material (in this case, water) outwards. The quantities ET and P represent total evapotranspiration and precipitation from/to the ground surface S.

In general, the lateral water flux can be positive at some height, and negative elsewhere. For example, one could have moisture influx at lower levels, and export or no flux at higher levels. Particular attention is required by the topping surface: if it is placed at the top of the atmosphere, there is no flux, if it is placed at the Tropopause (as often is), then one has to take into account the export of water at that level.

Forgetting, at least for now, the flux at the column top (which usually carries water away, if anything), an important issue concerns the recycling ratio between the water coming from local evapotranspiration inside the column (often due to precipitation recycling) and the water entering the column from the lateral boundaries (transport from elsewhere, for example from a nearby ocean) (Eltahir and Bras, 1996). Figure 4 illustrates this point.

The recycling ratio depends on the size of the region considered. If we consider a very small area, the recycling ratio can become very small. If we consider the whole Earth, the recycling ratio can become infinite, because there is no transport from outside (if we discard water brought to Earth by comets). So the recycling ratio is scale-dependent. But it is also dependent on geography: in regions such as the Amazon, recycling of precipitation is quite important, while in regions heavily affected by transport of moisture of marine origin it can become irrelevant.

The above considerations led some authors to devise a simplified model for describing the insurgence of summer droughts at continental midlatitudes (D'Andrea et al., 2006). The motivation for the modeling exercise was that the intense heat wave experienced by a wide portion of Europe in summer 2003 was associated with a low inflow of moisture from the Atlantic ocean, accompanied by a dry soil moisture anomaly at the beginning of summer. The dry soil conditions reduced evapotranspiration from the soil, further reducing the input of moisture in the atmosphere. The models discussed by D'Andrea et al. (2006) and Baudena et al. (2008) are based on a system of Ordinary Differential Equations describing the dynamics of either 4 or 5 state variables, namely atmospheric potential temperature and moisture averaged over the Planetary Boundary Layer (approximately, the

Figure 4. Schematic illustration of the water balance in a portion of the atmosphere.

lowest kilometer of atmosphere above the ground), the soil temperature and moisture averaged over the upper meter of soil, and, in a later version, the fractional vegetation cover (that is, the fraction of land covered with vegetation). Such variables provide a bulk description of the soil-vegetation-atmosphere dynamics in a large (almost continental-scale) box including the upper soil and the lower atmosphere.

That system, whose details can be found in the cited papers and are not worth reporting again here, was driven by an external parameter representing the lateral convergence (influx) of moisture from the ocean, and was evolved forward in time starting from different initial values of soil moisture. The results indicated that when the external water input is too low, the system always ends up in a dry and hot state (the "heat wave"), while in case of a large enough lateral water input, the system reaches a state of cooler (albeit warm) and rainy summer (the "standard conditions"). But in an intermediate range of lateral water inflow around 1 mm/day (that is, 1 kg/day/m^2), which is a rather normal value for the summer moisture inflow from the Atlantic Ocean, the system displays bistability, with one "hot and dry" state and one "warm and wet" state. The initial conditions on soil moisture determine which state is reached: dry soils at the beginning of summer lead to the hot and dry state, while larger initial values of soil moisture lead to the warm and wet state, in keeping with the fact that heat waves seem to require both a low inflow of water and a dry soil anomaly. Including the dynamics of vegetation added a new interesting effect: since transpiration from vegetation is larger than evaporation from bare soil, when the vegetation cover was below a certain threshold (which depends on the characteristics of the vegetation itself) the system ended on the dry state, because not enough water was transferred from the soil to the atmosphere (Baudena et al., 2008).

The simple model mentioned above was late extended to consider the case of a sandy planet, in which water is contained in the sand (Cresto Aleina et al., 2013). The question was, in that case, whether the presence of vegetation could trigger enough transpiration as to generate a hydrological cycle which sustains the vegetation itself. Using a super-simplified model with ten variables (temperature and moisture in two soil layers, potential temperature and moisture in the PBL and in the free Troposphere, vegetation cover and the liquid water content in the Troposphere) representing the whole planet (thus, no lateral transport), Cresto Aleina et al. (2013) were able to show that under a wide range of conditions the presence of vegetation does generate a sustained hydrological cycle and the planet remains vegetated. Although too simple to be taken too seriously, this simple model provides an example of how living organisms can generate their own niche, that is, how they can change the physical environment in a way that allows their survival.

6 Downscaling, upscaling and of the like

To provide unambiguous analysis of the changes and uncertainties in the global water cycle we require global data. These account for long-term in-situ observations and station-based gridded datasets, satellite data, as well as the outputs of numerical models of the climate system. The models, in particular, are essential to improve our understanding of the physical processes that drive the water cycle and the climate system in general, and to identify the main mechanisms at play and the feedbacks that connect small to large scales. Climate models are the most advanced tools that are currently available to make projections on the future evolution of the climate system under different emission and land-use scenarios. However, there is a scale mismatch between the spatial resolution that is currently achieved by the state-of-the-art GCMs or Earth System Models (at best of the order of 70-100 km) and the smaller scales at which the climate change impacts on ecosystems and hydrology, and related risks mostly act, and that are essential to provide a reliable representation of surface processes and atmosphere-surface interactions.

To fill this scale gap, a climate downscaling modelling chain is often adopted. In the first step of the chain, corresponding to the largest scales, Global Climate Models provide climate scenarios for the whole planet at relatively coarse spatial resolutions. The current trend of increasing the resolutions of GCMs is limited by the enormous computational and storage resources required, the limits of the physical approximations and the constrain of re-tuning the models whenever the resolution is increased. At intermediate scales, Regional Climate Models (RCMs) are nested into the GCMs which provide the appropriate boundary conditions (dynamical downscaling) to produce a higher-resolution estimate of the climatic variables. RCMs are used to simulate the Earth's climate system at a higher spatial resolution then the GCMs, but over a limited area. The finest resolution that can be commonly achieved with a regional model is about ten to twenty km, a scale still too large for obtaining data directly usable in hydrological models and for impact studies, e.g., to produce scenarios at the scale useful for water resource assessments and management. The use of higher-resolution nonhydrostatic RCMs in a climatic context would allow obtaining climate information, scenarios and projections at significantly finer resolutions (down to 3 km), but the computational effort required by this approach is rather formidable. Meteo-climatic forcing can be obtained at the relevant scale by statistical and stochastic downscaling of both GCMs and RCMs.

In the statistical downscaling approach one seeks to derive quantitative mathematical relations between large-scale predictors and local climate, by using transfer functions generally derived from observations. These cross-scale relationships are then applied to a GCM/RCM data to derive local-scale information consistent with the synoptic-scale forcing of the model and then to develop local climate change scenarios (see e.g. Maraun et al., 2010). Another effective approach, specifically devised to downscale precipitation fields, relies on stochastic downscaling methods. Stochastic rainfall downscaling aims at generating synthetic high-resolution precipitation data, whose statistical properties are consistent with the small-scale statistics of the observed (e.g., radar) precipitation, based only on the knowledge of the large-scale precipitation field, either produced by a GCM or RCM or derived from coarse-scale observations (for example, satellite measurements). Among the many stochastic downscaling methods available, a particularly flexible procedure is the Rainfall Filtered Autoregressive Model (RainFARM; Rebora et al., 2006; D'Onofrio et al., 2014). Further down along the modelling chain, the downscaled model outputs can be used to force hydrological models for rainfall/runoff and groundwater flow to simulate the hydrological response at the scales of interest (Sorooshian et al., 2008) and produce outputs that are eventually usable, at the very end of the chain, by impact modellers.

In addition to this kind of applications, it is worth specifying that stochastically downscaled model fields can be used, instead of fields interpolated in a smooth way, to force land surface schemes, which is a good strategy for obtaining a better representation of the heat and moisture fluxes from the soil and vegetation to the atmosphere. Then, in order to correctly represent the land surface-atmosphere interaction, climate upscaling methods and new parameterizations of the small-scale processes occurring at the surface have to be developed.

In the last few years, the reliability of both climate models and downscaling techniques has improved (Rahmstorf et al., 2007; Reichler and Kim, 2008) and a few applications already exist in which the downscaled model outputs have been used to drive hydrological models and predict changes in water resources availability (see, among others, Barnett and Pierce, 2008; Chenoweth et al., 2011). However, the cascade of uncertainty from climate to hydrological and impact modelling across the downscaling chain has not yet been taken into account in a systematic way. Each element of the chain is equipped with uncertainties that should be considered, quantified, and clearly communicated. At this point, a few questions arise:

- How do uncertainties propagate across the chain, from large to small scales, and how we can provide an estimate of the final uncertainty?
- Do the uncertainties associated with small-scale (1 km and less) dynamics at the surface, soil and vegetation feed back on to the atmosphere and how?
- Can methodologies be developed and applied in order to reduce uncertainties?
- What is the role of the observations (and data assimilation) along the chain?

Uncertainties inherent in the climate and hydrological models can be reduced by increasing our knowledge of the functioning of the system, which can be achieved using different strategies depending upon the model type and the considered process. Epistemic uncertainty in hydrological modeling, for example, can be reduced using detailed distributed models coupled with field-scale measurements of the state variables. The observations are important elements in many steps of the chain: for calibrations, verification, testing, and for validation purposes. For example, the GCM/RCM outputs and their downscaled fields on historical periods can be compared with the available measurements, in order to evaluate the performances of the models and downscaling procedures, and then used in hydrological simulations for validation against measured data (e.g., streamflow), in order to validate the modeling chain. New strategies can be adopted in which stochastic precipitation downscaling is applied to model precipitation fields before their usage to force vegetation and soil dynamics in order to verify whether this approach leads to improved estimates of the land surface fluxes, again, by comparing with the data from field studies.

In the last decade the number of studies focusing on the impact of climate change on the hydrological cycle increased enormously and this can now be considered as a rapid growing subject at the border between climatic and hydrological sciences. A modelling framework that can operate both in downscaling and upscaling modes would provide the necessary capabilities for addressing scaling issues, coupling of feedbacks at multiple scales, and finally understanding and predicting changes in the global water cycle.

7 Concluding remarks

Water - in liquid, vapor or solid form - is at the heart of the Earth's climate system. A planet with a dry atmosphere (and a dry thermodynamics without water phase changes) would be completely different. And water is a prototype for a fluid, able to flow between different regions and components of the climate machine.

Water, as a polar solvent, allows the presence of life on our planet. Indeed, current theoretical estimates of planetary habitability are built upon the search for the physical conditions where water can exist in liquid form (Kasting and Catling, 2003; Seager, 2013). What makes our planet special (apart from the fact that we live on it) is the presence of water in its three forms, the presence of life and the active geodynamics of its interior, with the multi-scale biogeochemical cycles that couple atmosphere, oceans, surface and Earth interior.

Water is crucial for human societies, as it is dramatically shown by the effects of droughts (when water is scarce) or floods (when water is too much). Water is needed for drinking purposes, for agriculture, for industry, and for hydropower production. Assessing and predicting current and future changes in the many components of the hydrological cycle is a hard task, but one that bears the most important consequences for adaptation and risk mitigation strategies.

Bibliography

Archer, D. R., H. J. Fowler, *Spatial and temporal variations in precipitation in the upper Indus basin, global teleconnections and hydrological implications*, Hydrol. Earth Syst. Sci., **8**, 47-61, doi:10.5194/hess-8-47-2004, 2004.

Ashkenazy, Y., Y. Feliks, H. Gildor, and E. Tziperman, *Asymmetry of daily temperature records*, J. Atmos. Sci. **65**, 3327-3336, 2008.

Barnett, T. P., D. W. Pierce, *When will Lake Mead go dry?*, Water Resour. Res., **44**, W03201, doi:10.1029/2007WR006704, 2008.

Baudena, M., F. D'Andrea, A. Provenzale, *A model for soil-vegetation-atmosphere interactions in water-limited ecosystems*, Water Resour. Res., **44**, W12429, doi:10.1029/2008WR007172, 2008.

Beniston, M., *Climate change and its impacts: growing stress factors for human societies*, International Review of the Red Cross, **92**, 557-568, doi:10.1017/S1816383110000342, 2010.

Bookhagen, B., D. W. Burbank, *Toward a complete Himalayan hydrological budget: Spatiotemporal distribution of snowmelt and rainfall and their impact on river discharge*, J. Geophys. Res., **115**, F03019, doi:10.1029/2009JF001426, 2010.

Chenoweth, J., P. Hadjinicolaou, A. Bruggeman, J. Lelieveld, Z. Levin, M. A. Lange, E. Xoplaki, M. Hadjikakou, *Impact of climate change on the water resources of the eastern Mediterranean and Middle East region: Modeled 21st century changes and implications*, Water Resour. Res., **47**, W06506, doi:10.1029/2010WR010269, 2011.

Cresto Aleina, F., M. Baudena, F. D'Andrea, A. Provenzale, *Multiple equilibria on planet Dune: climate-vegetation dynamics on a sandy planet*, Tellus B, **65**, 17662, doi:tellusb.v65i0.17662, 2013.

Dijkstra, H. A., *On the interaction of SST modes in the North Atlantic Ocean*, J. Phys. Oceanogr., **36**, 286-299, 2006.

D'Andrea, F., A. Provenzale, R. Vautard, N. de Noblet-Decoudré, *Hot and cool summers: Multiple equilibria of the continental water cycle*, Geophys. Research Lett., **33**, L24807, doi:10.1029/2006GL027972, 2006.

D'Onofrio, D., E. Palazzi, J. von Hardenberg, A. Provenzale, S. Calmanti, *Stochastic Rainfall Downscaling of Climate Models*, J. Hydrometeor, **15**, 830-843, doi: http://dx.doi.org/10.1175/JHM-D-13-096.1, 2014.

E. A. B. Eltahir, R. L. Bras, *Precipitation Recycling*, Rev. Geophysics, **34**, 367-378, 1996.

Filippi, L., E. Palazzi, J. von Hardenberg, A. Provenzale, *Multidecadal Variations in the Relationship between the NAO and Winter Precipitation in the Hindu KushKarakoram*, J. Climate, **27**, 7890-7902, doi: http://dx.doi.org/10.1175/JCLI-D-14-00286.1, 2014.

Giorgi, F., E.-S. Im, E. Coppola, N. S. Diffenbaugh, X. J. Gao, L. Mariotti, Y. Shi, *Higher hydroclimatic intensity with global warming*, J. Climate, **24**, 5309-5324, doi:10.1175/2011JCLI3979.1, 2011.

Hazeleger, W., X. Wang, C. Severijns, S. tefănescu, R. Bintanja, A. Sterl, K. Wyser, T. Semmler, S. Yang, B. van den Hurk, T. van Noije, E. van der Linden, K. van der Wiel, *EC-Earth v2.2: description and validation of a new seamless earth system prediction model*, Clim. Dyn., **39**, 2611-2629, doi:10.1007/s00382-011-1228-5, 2012.

Held, I. M., B. J. oden, *Water vapor feedback and global warming*, Annu. Rev. Energy Environ., **25**, 441-475, doi:110.1146/annurev.energy.25.1.441, 2000.

Hoerling, M., J. Eischeid, J. Perlwitz, X. Quan, T. Zhang, P. Pegion, *On the Increased Frequency of Mediterranean Drought*, J. Climate, **25**, 2146-2161, doi:10.1175/JCLI-D-11-00296.1, 2012.

IPCC, *Climate Change 2013: The Physical Science Basis. Contribution of Working Group I to the Fifth Assessment Report of the Intergovernmental Panel on Climate Change* [Stocker, T.F., D. Qin, G.-K. Plattner, M. Tignor, S.K. Allen, J. Boschung, A. Nauels, Y. Xia, V. Bex and P.M. Midgley (eds.)], Cambridge University Press, Cambridge, United Kingdom and New York, NY, USA, 1535 pp, 2013.

IPCC, *Climate Change 2014: Impacts, Adaptation, and Vulnerability. Part A: Global and Sectoral Aspects. Contribution of Working Group II to the Fifth Assessment Report of the Intergovernmental Panel on Climate Change* [Field, C.B., V.R. Barros, D.J. Dokken, K.J. Mach, M.D. Mastrandrea, T.E. Bilir, M. Chatterjee, K.L. Ebi, Y.O. Estrada, R.C. Genova, B. Girma, E.S. Kissel, A.N. Levy, S. MacCracken, P.R. Mastrandrea, and L.L. White (eds.)], Cambridge University Press, Cambridge, United Kingdom and New York, NY, USA, 1132 pp., 2014.

Jasechko S., Z. D. Sharp, J. J. Gibson, S. J. Birks, Y. Yi, P. J. Fawcett, *Terrestrial water fluxes dominated by transpiration*, Nature, **496**, 347-350, doi:10.1038/nature11983, 2013.

Kasting J. F, D. Catling, *Evolution of a Habitable Planet*, Annual Review of Astronomy and Astrophysics, **41**, 429-463, doi:10.1146/annurev.astro.41.071601.170049, 2003.

Kaufman Y. J., R. S. Fraser, *The effect of smoke particles on clouds and climate forcing*, Science, **277**, 1636-1639, doi:10.1126/science.277.5332.1636, 1997.

Lal M., G. A. Meehl, J. M. Arblaster, *Simulation of Indian summer monsoon rainfall and its intraseasonal variability in the NCAR climate system model*, Reg. Environ. Change., **1**, 163-179, doi:10.1007/s101130000017, 2000.

Lee J. Y., B. Wang, I. S. Kang, J. Shukla, A. Kumar, J. S. Kug, J. K. E. Schemm, J. J. Luo, T. Yamagata, X. Fu, O. Alves, B. Stern, T. Rosati, C. K. Park, *How are seasonal prediction skills related to models' performance on mean state and annual cycle?* Clim. Dyn., **35**, 183-267, doi:10.1007/s00382-010-0857-4, 2010

McCarthy, M. P, P. W. Thorne, H. A. Titchner, *An Analysis of Tropospheric Humidity Trends from Radiosondes*, J. Climate, **22**, 5820-5838, doi:10.1175/2009JCLI2879.1, 2009.

Maraun, D., F. Wetterhall, A. M. Ireson, R. E. Chandler, E. J. Kendon, M. Widmann, S. Brienen, H. W. Rust, T. Sauter, M. Themeßl, V. K. C. Venema, K. P. Chun, C. M. Goodess, R. G. Jones, C. Onof, M. Vrac, I. Thiele-Eich, *Precipitation downscaling under climate change: Recent developments to bridge the gap between dynamical models and the end user*, Rev. Geophys., **48**, RG3003, doi:10.1029/2009RG000314, 2008.

Margari, V., L. C. Skinner, P. C. Tzedakis, A. Ganopolski, M. Vautravers, N. J. Shackleton, *The nature of millennial-scale climate variability during the past two glacial periods*. Nature Geoscience, **3**, 127-133, doi:10.1038/NGEO740, 2010.

Palazzi, E., J. von Hardenberg, A. Provenzale, *Precipitation in the Hindu-Kush Karakoram Himalaya: Observations and future scenarios*, J. Geophys. Res. Atmos., **118**, 85-100, doi: 10.1029/2012JD018697, 2013.

Palazzi, E., J. von Hardenberg, S. Terzago, A. Provenzale, *Precipitation in the Karakoram-Himalaya: a CMIP5 view*, Clim. Dyn., **45(1-2)**, 21-45, doi: 10.1007/s00382-014-2341-z, 2015.

Pieri, A., J. von Hardenberg, A. Parodi, A. Provenzale, *Sensitivity of precipitation statistics to resolution, microphysics and convective parameterization: a case study with the high-resolution WRF climate model over Europe*, J. Hydrometeor., **16**, 1857-1872, doi:10.1175/JHM-D-14-0221.1, 2015.

Rahmstorf, S., A. Cazenave, J. A. Church, J. E. Hansen, R. F. Keeling, D. E. Parker, R. C. J. Somerville, *Recent Climate Observations Compared to Projections*, Science, **316(5825)**, p. 709, [DOI:10.1126/science.1136843, 2007.

Rasmussen R., B. Baker, J. Kochendorfer, T. Meyers, S. Landolt,A. P. Fischer, J. Black, J. M. Thériault, P. Kucera, D. Gochis, C. Smith, R. Nitu, M. Hall, K. Ikeda, E. Gutmann, *How well are we measuring snow: the*

NOAA/FAA/NCAR winter precipitation test bed, Bull. Am. Meteorol. Soc., **93**, 811-829, doi:10.1175/BAMS-D-11-00052.1, 2012.

Rebora, N., L. Ferraris, J. von Hardenberg, A. Provenzale, *RainFARM: Rainfall downscaling by a filtered autoregressive model*, J. Hydrometeor., **7**, 724-738, doi:10.1175/JHM517.1, 2006.

Reichler, T., J. Kim, *How Well Do Coupled Models Simulate Today's Climate?*, Bull. Amer. Meteor. Soc., **89**, 303-311, doi: 10.1175/BAMS-89-3-303, 2008.

Rietkerk, M., V. Brovkin, P. M. van Bodegom, M. Claussen, S. C. Dekker, H. A. Dijkstra, S. V. Goryachkin, P. Kabat, E. H. van Nes, A. -M. Neutel, S. E. Nicholson, C. Nobre, V. Petoukhov, A. Provenzale, M. Scheffer, S. I. Seneviratne, *Local ecosystem feedbacks and critical transitions in the climate*, Ecological Complexity, **8**, 223-228, doi:10.1016/j.ecocom.2011.03.001, 2011

Schneider, U., A. Becker, P. Finger, A. Meyer-Christoffer, M. Ziese, B. Rudolf, *GPCC's new land surface precipitation climatology based on quality-controlled in situ data and its role in quantifying the global water cycle*, Theor. Appl. Climatol., **115(1-2)**, 15-40, 2014.

Seager, S., *Exoplanet Habitability*, Science, **3**, 577-581, doi:10.1126/science.1232226, 2013.

Sherwood, S. C., C. L. Meyer, *The general circulation and robust relative humidity*, J. Clim., **19**, 6278-6290, doi: 10.1175/JCLI3979.1, 2006.

Sorooshian, S., K. -l. Hsu, E. Coppola, B. Tomassetti, M. Verdecchia, G. Visconti (Eds.), *Hydrological Modelling and the Water Cycle. Coupling the Atmospheric and Hydrological Models*, Series: Water Science and Technology Library, **63**, 291 p., doi: 10.1007/978-3-540-77843-1, 2008.

Spinoni, J., G. Naumann, J. Vogt, P. Barbosa, *European drought climatologies and trends based on a multi-indicator approach*, Global and Planetary Change, **127**, 50-57, doi: 10.1016/j.gloplacha.2015.01.012, 2015.

Syed, F. S., F. Giorgi, J. S. Pal, M. P. King, *Effect of remote forcings on the winter precipitation of central southwest Asia part 1: observations*, Theor. Appl. Climatol., **86(1-4)**, 147-160, doi:10.1007/s00704-005-0217-1, 2006.

Syed, F. S., F. Giorgi, J. S. Pal, K. Keay, *Regional climate model simulation of winter climate over central-southwest Asia, with emphasis on NAO and ENSO effects* Int. J. Climatol., **30**, 220-235, doi:10.1002/joc.1887, 2010.

Su F., X. Duan, D. Chen, Z. Hao, L. Cuo, *Evaluation of the global climate models in the CMIP5 over the Tibetan Plateau*, J. Clim.,**26**, 3187-3208. doi:10.1175/JCLI-D-12-00321., 2012

Tebaldi C., R. Knutti, *The use of the multi-model ensemble in probabilistic climate projections*, Phil. Trans. R. Soc. Lond. A, **365**, 2053-2075, 2007.

Trenberth, K. E.,, *Changes in precipitation with climate change*, Clim. Res., **47**, 123-1385, 2011.

Turco, M., E. Palazzi, J. von Hardenberg, and A. Provenzale, *Observed climate change hotspots*, Geophys. Res. Lett., textbf42, 3521-3528. doi: 10.1002/2015GL063891, 2015.

Wang, Y. -H., G. Magnusdottir, H. Stern, X. Tian, Y. Yu, *Decadal variability of the NAO: Introducing an augmented NAO index*, Geophys. Res. Lett., **39**, L21702, doi:10.1029/2012gl053413, 2012.

Yadav R. K., K. R. Kumar, M. Rajeevan, *Increasing influence of ENSO and decreasing influence of AO/NAO in the recent decades over northwest India winter precipitation*, J. Geophys. Res., **114**, D12112, doi:10.1029/2008JD011318, 2009.

Yadav R. K., J. H. Yoo, F. Kucharski, M. A. Abid, *Why Is ENSO Influencing Northwest India Winter Precipitation in Recent Decades?*, J. Climate., **23**, 1979-1993, doi:10.1175/2009JCLI3202.1, 2010.

Yadav R. K., K. R. Kumar, M. Rajeevan, *Characteristic features of winter precipitation and its variability over northwest India*, J. Earth Syst. Sci., **121**, 611-623, doi:10.1007/s12040-012-0184-8, 2012.

Climate dynamics on watershed scale: along the rainfall-runoff chain

Klaus Fraedrich[†], Frank Sielmann[‡], Danlu Cai[*], and Xiuhua Zhu[‡]
[†] Max Planck Institute of Meteorology, Hamburg, Germany
[‡] University of Hamburg, Hamburg, Germany
[*] Institute of Remote Sensing and Digital Earth, Chinese Academy of Sciences, Beijing, China

Abstract Climate regimes of watersheds are determined by the rainfall-runoff chain, which comprises processes of a wide range of time and space scales. They are presented here in geographical and eco-hydrological state spaces combining observations and minimalist concepts. (*i*) Water supply and demand govern the rainfall-runoff chain: Rainfall provides the water supply, net radiation represents the water demand which, related to water supply, separates water from energy limited regimes and determines the discharge as the probability of rainfall reaching the soil water reservoir. This macro-state of the rainfall-runoff chain is described by an empirical equation of state and derived by a (micro-state) biased coin-flip Ansatz. (*ii*) Eco-hydrologic spaces spanned by water and energy fluxes or flux ratios, embed the states of the rainfall-runoff chain as occurrence probabilities and their time changes as trajectories. Tracers, like vegetation-greenness, can also be included to estimate watershed resilience and, based on the trajectories, to attribute the changes to external (or climate) and internal (or anthropogenic) causes. (*iii*) Geo-morphological patterns like the area ratios of closed lakes (lake/basin), soil moisture storage of watersheds, and drainage densities of river systems, can be functionally related to eco-hydrological climates. Suitable minimalist models of the rainfall-runoff chain provide estimates of regional climate regimes governing the watersheds of the past, present, or future.

1 Introduction

Analyzing, modeling, understanding and applying the dynamics of complex processes which govern the Earth's environment can be described in a set of spaces: A space of complexity captures the required or envisaged details of understanding, modeling and applications, which range from comprehensive

A. Provenzale et al. (Eds.), *The Fluid Dynamics of Climate*,
CISM International Centre for Mechanical Sciences
DOI 10.1007/ 978-3-7091-1893-1_8

to conceptual – depending on the degree of freedom – that is, for example, from state of the art global climate models versus minimalist toys (Fraedrich et al., 2005). In this sense this is an analogy to Popper's (1972) complexity spectrum described in the essay "Of Clocks and Clouds" commonly related to the essence of predictability and probability (see Fraedrich, 2007). Here we are also following a minimalist path for eco-hydrological modeling and diagnostics on watershed scale in order to complete the "climate dynamics on the global scale" (Fraedrich et al., 2015c) contributing to analyses on resilience, hysteresis and the attribution of climate change. In the geographic space the measured reality is presented displaying climate or sequences of the weather evolution as observed by in situ and remote sensors or simulated. Physical state spaces are spanned by the measured or simulated physical variables describing the dynamics of the eco-hydrological processes. And, finally, the link to the human dimension possibly connects the quantitative assessment of uncertainties (of climate models, Holden et al., 2014) with quantitative estimates of global risk measures including estimated uncertainties (Beese et al., 1998).

The aim of this chapter[1] is to present – in a comprehensive way – results and novel interpretations of the eco-hydrological dynamics on watershed scale and presenting it in the geographical and physical spaces utilizing a minimalist modeling and analysis approach. First, the ideal rainfall-runoff chain is introduced, which communicates between water supply and demand satisfying water and energy budgets and an equation of state (Section 2, see Fraedrich, 2010). Secondly, the eco-hydrological state spaces, their variables and diagrams are presented including first analyses of geobotanic classifications, a validation of the parsimonious concept of the rainfall-runoff chain and its application to sensitivity (or elasticity) analyses (Section 3, Fraedrich and Sielmann, 2011; Fraedrich et al., 2013, 2015b). Finally, geomorphological patterns related to vegetation, rivers, soil, and lakes are analyzed, conceptually modeled, and applied following the parsimonious approach described (Section 4, Cai et al., 2014; Fraedrich, 2015a). The conclusion provides a brief outlook to the human dimension (Section 5).

2 Ideal rainfall-runoff chain between water supply and demand

Land surface climates on catchment scale are controlled by long term means and area averaged water and energy fluxes satisfying the respective flux

[1] A summary of our recent research

balance equations (defined by water equivalents of energy units)

$$\text{Water flux balance} \qquad 0 = P - Ro - E \tag{1}$$

$$\text{Energy flux balance} \qquad 0 = N - H - E \tag{2}$$

Precipitation P and net radiation N are forcing terms representing water supply and demand. The balancing fluxes are subjected to processes which, partitioning runoff Ro plus evaporation E at the ground, and sensible heat H plus moisture fluxes to the atmosphere, E, comprise the rainfall-runoff chain and link atmosphere, biosphere and pedosphere. These climate variables are functionally related by an *equation of state* for ideal land surface climates that, not unlike the ideal gas law, has been discovered empirically (Schreiber, 1904) and represents physical and stochastic properties of the rainfall-runoff chain.

$$\text{Equation of state} \qquad Ro = P \exp\left(-N/P\right) \tag{3}$$

Equation 3 relates the discharge to the rainfall which has not been evaporated; that is, the probability of the water supply to reach the soil water reservoir. In this sense an increasing runoff-rainfall ratio W is linked to a decreasing dryness ratio $D = N/P$ (or increasing wetness ratio $1/D$) and vice versa.

The random process of daily rainfall P transferring water through the fast biospheric reservoir to the slow soil water reservoir appears to be a process analog to random molecules transferring momentum through reflection to a slow wall. While the former yields an equation of state for the ideal rainfall-runoff chain (empirically discovered more than a century ago), the latter has led to the ideal gas law (empirically discovered long before theoretically substantiated). The theory (Fraedrich, 2010) underlying Schreiber's (1904) empirical rainfall-runoff chain in equilibrium is based on the following concept: The rainfall chain commences from a fast stochastic water reservoir of limited small capacity representing water intercepted in vegetation, mulch, etc., which feeds into a slow (almost stationary) soil moisture reservoir of large capacity balancing its runoff Ro after long-term averaging. The fast reservoir's water is supplied by rainfall, prescribed by a maximum entropy (exponential) distribution determined only by its expected value P. Its capacity is defined by (the water equivalent of) net radiation N, which is available to evaporate water. Thus, if rainfall entering the fast biospheric reservoir is "larger" or "smaller" than its capacity, a biased coin-flip statistics establishes the following day-to-day occurrence probability for water surplus: precipitation larger than capacity enters the slow reservoir while the residual evaporates so that the fast reservoir can re-start as "empty".

Rainfall below capacity evaporates completely and the residual available energy provides sensible heat H to the atmosphere; now the fast reservoir can re-start again as "empty". The water surplus for the slow (soil) reservoir integrated over the occurrence probability of the biased coin-flip provides the climate mean discharge as given by Schreiber's empirical formula (Eq. 3).

3 Eco-hydrological state spaces: Variables and Diagrams

Water and energy fluxes and their combination to non-dimensional flux-ratios characterize, not unlike the similarity measures in fluid dynamics, the climate mean state of a watershed. Pairs of these quantities (fluxes and flux-ratios) enumerated below are used to span two-dimensional spaces to embed the (climate mean) states of continental watersheds and their changes in terms of trajectories.

3.1 Fluxes and flux ratios

The following flux ratios and their functional relationships provided by the water and energy balance and equation of state (Eqs. 1 to 3) are displayed in ecohydrological state spaces (Fig. 1) spanned by the flux ratios.

Dryness ratio (aridity index): Dryness or aridity combines energy and water fluxes into one parameter relating water demand (energy supply) to water supply and it separates water from energy limited regimes at $D = 1$. Following Budyko (1974) the dryness ratio (aridity index) characterizes geobotanic states of watersheds:

$$D = N/P$$

Runoff-rainfall ratio (or excess water): Relating water fluxes to their supply by rainfall yields the runoff ratio

$$W = Ro/P$$

It describes the streamflow of rivers on a regional or catchment scale, characterizes the natural rainfall-runoff chain in terms of relative excess rainfall $(P - E)/P$, and represents the fraction of water unused by an ecosystem (but available for terrain formation, Milne et al., 2001) when compared to the total water supply. In this sense the runoff-rainfall ratio is also a measure of the water efficiency of an ecosystem. Negative excess rainfall occurs if evaporation is supported by hydrologic and not by atmospheric processes. With the equation of state we obtain $W = \exp(-D)$. Note that the evaporation ratio $F = 1 - W$ is more frequently used in hydrological analyses and both ratios are easily envisaged in Fig. 1a.

Energy flux ratio (or excess energy): Relating energy fluxes to their supply yields a relative energy flux ratio

$$U = H/N$$

which describes the fraction of energy unused by the system when compared to its supply. That is, the relative excess energy $(N - E)/N$ (available for photosynthesis, Milne et al., 2001) which, not unlike the water efficiency W, is a measure of the energy efficiency of an ecosystem. Both water and energy excess (or efficiency) describe proportions of available water and energy which, remaining unused, appear to be relevant to identify the causes of climate and basin change. Note that U and W determine the dryness ratio, $D = (1 - W)/(1 - U)$ and, the equation of state (Eq. 3) can be formulated as $U = 1+(1-W)/\ln(W)$, with the slope of change $(dU/dW) = -\{\exp(D) - (D+1)\}/D^2$.

Bowen-ratio: The Bowen ratio describes the latent-versus-sensible heat flux ratio connecting water and energy flux balances, and $B = H/E = UD/(1-W)$. Introducing Schreiber's formula, the steady state energy flux balance $E = N/(1 + B)$ provides a Bowen ratio $B = D/F = D/\{1 - \exp(-D)\} - 1$.

Geobotanic zonality and meridionality: The Earth surface climate states are characterized by watershed scale eco-hydrologic regimes in terms of the mean fluxes and/or flux ratios satisfying the rainfall-runoff chain (Eqs. 1 to 3), which are generated by microstate fluctuations of rainfall, modulated by the biospheric interception, and averaged over the timescale of the slow soil reservoir. A set of essential state space diagrams is introduced to display and analyze eco-hydrological states and their dynamics. In Budyko's framework, climate states can be embedded in a two-dimensional state space spanned by net radiation N (representing the meridionality of zonal the mean energy forcing; see Fig. 2) and dryness D, which comprises the climate mean sources of the rainfall-runoff chain, that is water demand versus supply:

(i) The state space displays climate variables such as precipitation, presented by the slopes of isolines through the origin, $N = PD$, and the

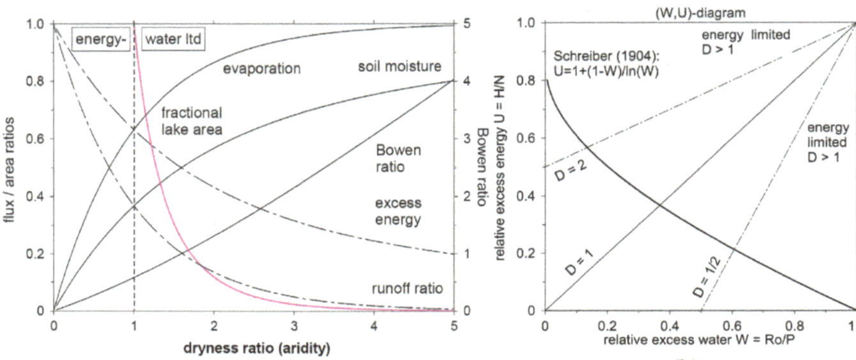

(a) (b)

Figure 1. Eco-hydrologic state space spanned by ratios of mean fluxes functionally related by the stochastic rainfall-runoff chain (equation of state): (a) Bowen ratio B, runoff ratio W (excess water), and evaporation ratio F depending on the dryness index D. The lake area ratio A and soil moisture ratio are also introduced. (b) (W, U)-diagram spanned by excess water $W = Ro/P$ (runoff-rainfall ratio) and excess energy $U = H/N$, displaying the dryness ratio D and the ideal rainfall-runoff chain (equation of state, Eq. 3).

isolines of total supply, $N = (P + N)D/(D - 1)$, attain the slope of $dN/dD = P + N$ at the origin ($N = D = 0$).

(ii) This state space has also been used to include the biosphere and vegetation-based climate classification into a physics-based framework of climate analysis. Interpreting the energy input N as potential evapotranspiration, the dryness or aridity index D represents the flux ratio of water demand to supply. This ratio also provides a quantitative geobotanic measure of the climate-vegetation relation (Budyko, 1974): Tundra, $D < 1/3$, and forests, $1/3 < D < 1$, are energy limited because available energy N is low, so that runoff exceeds evaporation for given precipitation, $E \sim N$. Steppe and Savanna, $1 < D < 2.0$, semi-desert $2.0 < D < 3.0$, and desert $3.0 < D$, are water limited climates, where the available energy is so high that water supplied by precipitation evaporates, which then exceeds runoff.

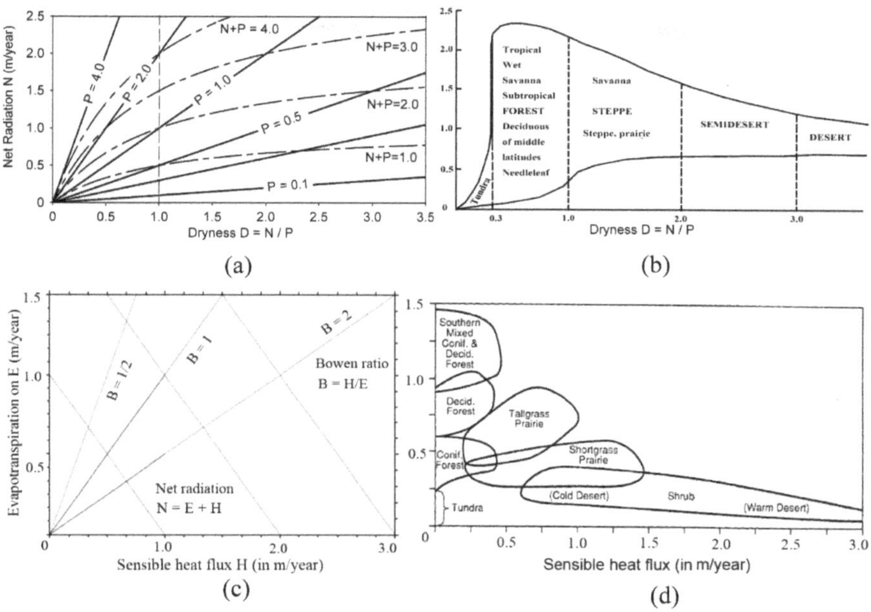

Figure 2. State space diagrams of the surface climate; the (N, D)-space (Budyko, 1974) is spanned by net radiation N (or potential evapotranspiration) and dryness ratio D: (a) Precipitation, $N = P \cdot D$, corresponds to the slope of straight lines through the origin; the total supply of energy and water curve satisfies $N = (N + P) \cdot D/(D + 1)$, all units in m/year water equivalent. (b) Geobotanic types are adapted from Budyko (1974, Figure 103); the boundary (full line) includes all area-units of (N, D)-pairs observed on the global land surface; dryness or D-classes are separated by vertical lines (dashed) with additional sub-classes depending on the magnitude of net radiation. The $D = 1$ threshold separating energy from water limited climate regimes (vertical line) is included. The (H/E)-diagrams (from Stephenson, 1990, Figure 3) spanned by sensible heat flux H and evapotranspiration E: (c) The slope of lines through the origin represent the inverse of the constant Bowen ratio $B = H/E$, while the off-diagonals present lines of constant net-radiation N (potential evapotranspiration) $E = N - H$; (d) Geobotanic types (Stephenson, 1990) represent N-American biomes, which are basically aligned along a latitude circle of about constant net-radiation.

3.2 Eco-hydrological diagnostics

Budyko's (1974) framework of analysis is first presented as a two-dimensional frequency distribution in the related eco-hydrological state space before being employed to validate the performance of the minimalist Ansatz (Eqs. 1 to 3) to represent the climate mean hydrological cycle on watershed scale. This is followed by sensitivity studies to interpret causal relations in a linear framework and, finally to introduce a diagnostics of the attribution of the causes of observed changes. The climates analyzed are based on ERA-Interim observations (25-yr averages: 1982-2006) and simulations by a state of the art GCM (present day IPCC4 scenario of 20th century control period 1958-2001, see Roeckner et al., 2006), which provide suitable datasets of regional land surface climate averages. Note that the simulations do not include direct human induced watershed and land surface changes.

State space distributions: The climatology is defined by the frequency distributions of the observed states in the (N, D)-diagram (for the global continents; Cai et al., 2014) and in the (W, U)-diagram for S-America. The global distribution (Fig. 3a) shows meridional change in terms of net radiation (vertical axis) increasing from small values at the pole to high values at the equator and the peaks are aligned along the $D = 1$ threshold separating water $(D > 1)$ from energy limited regime $(D < 1)$, which is related to tropical and mid-latitude forests, respectively. The S-America frequency distribution (Fig. 3b) of excess water versus excess energy in the (W, U)-diagram is very well aligned along the rainfall-runoff equation of state (Eq. 3) with its primary peak at the $(D = 1)$ threshold and the secondary occupying the energy limited regime $(D < 1)$.

Validation – consistency and predictability: The ideal rainfall-runoff chain is validated in two steps: (*i*) Consistency is assessed comparing the ideal rainfall-runoff chain with the coupled GCM simulation by sampling the simulated Koeppen classes (Koeppen, 1936, A to F) in bins of the simulated dryness ratio D. The sample averages and standard deviations (vertical and horizontal axes centered on the means) show that the discrete Koeppen climate types (Fig. 4a) are well aligned along the ideal rainfall-runoff chain's dryness dependent runoff ratio (Fig. 4b, the black line indicates the ideal rainfall-runoff chain of Eq. 3). The geobotanic dryness D and Koeppen classes may fit Schreiber's equation better after suitably regrouping the Koeppen climate classes (including the subclasses; see Hanasaki et al., 2008). (*ii*) Predictability is analyzed comparing the simulated runoff R with the runoff Ro^* derived by Schreiber's equation of state using simulated dry-

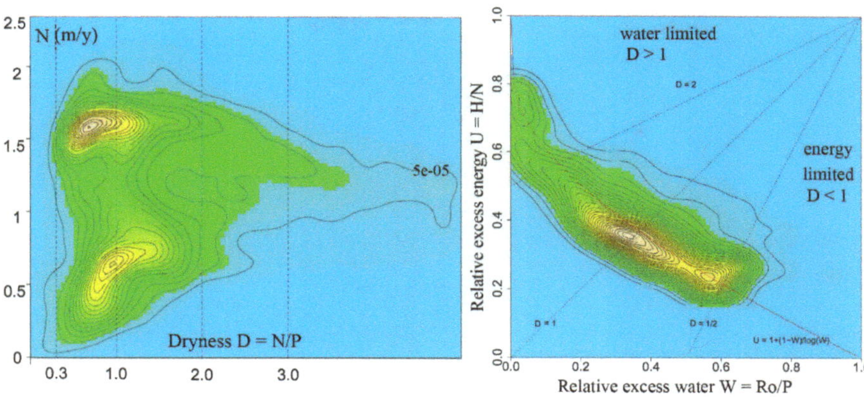

(a) (b)

Figure 3. Frequency distribution of the climates of land surface fluxes and flux ratios based on grid point from ERA-Interim (the total land area corresponds to the total number of equal area units): (a) (N, D)-pairs of net radiation N and dryness D for the global continents; the $D = 1$ threshold separating energy- from water-limited regimes (vertical line) is included. (b) (U, W)-pairs of energy excess U and water excess W for S-America. The dryness D isolines are included and the equation of state (Eq. 3).

ness ratio D and precipitation P (Fig. 4c,d). It is verified as demonstrated by the (Ro, Ro^*)-scatter plot. The consistency between GCM-simulated dryness D dependent runoff-rainfall ratios W is also presented sampled in W-bins with means and standard deviations (large dots and horizontal lines) being compared with Schreiber's formula (Eq. 3, dashed) in (W, D)-space. These results demonstrate the stochastic rainfall-runoff chain's consistency with the phenomenological (Koeppen) climate classes and its predictability (within a consistent data set) for runoff, excess rainfall or energy.

Sensitivity, elasticity, susceptibility: Sensitivity to small changes of boundary conditions is deduced, because it is of relevance for interpreting climate change estimates and model performance. Focusing on runoff changes, these are described to first order, $\Delta Ro = Ro_P \Delta P + Ro_N \Delta N$, by partial differentials, Ro_P and Ro_N. For example, the Schreiber-Budyko Ansatz as an ideal or reference rainfall-runoff chain yields $Ro_P = (1 + D)\exp(-D)$ and $Ro_N = -\exp(-D)$. Rearrangement shows that the runoff sensitivity, $\Delta Ro/Ro$, depends on the sensitivities of water supply, $\Delta P/P$,

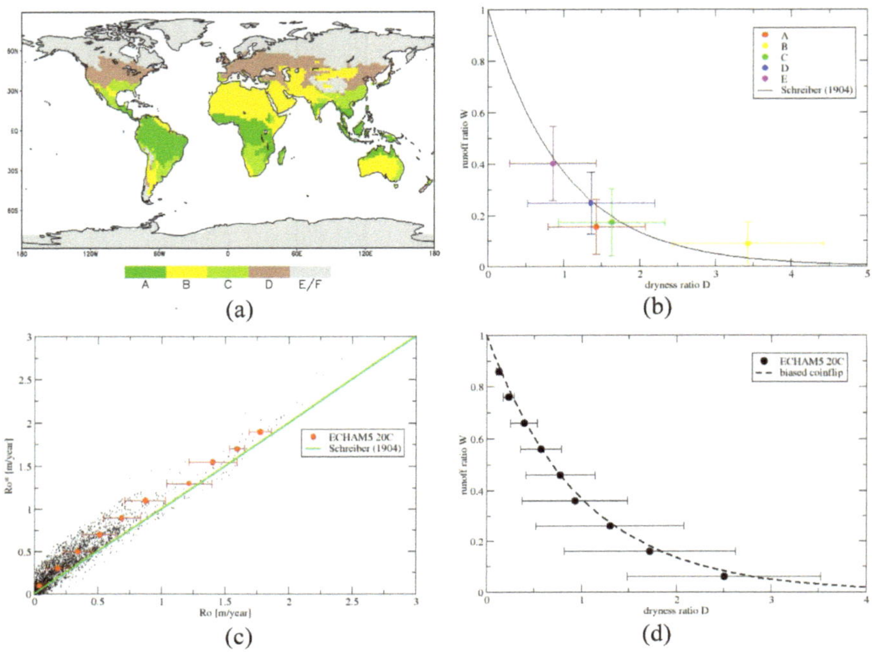

Figure 4. Validation of an ideal rainfall-runoff chain in a GCM environment with present day surface climates simulated: (a) Koeppen (1936) climates and (b) their locations in the idealized (W, D)-diagram (see Fig. 1a). (c) Ro^* derived by Schreiber's formula (using simulated dryness D and rainfall P) is compared with the GCM-simulated runoff Ro in a (Ro, Ro^*)-scatter plot. (d) GCM-simulated dryness D and excess rainfall W sampled in W-bins (means, standard deviations as dots, horizontal lines) follow Schreiber's formula (Eq. 3, dashed).

and demand, $\Delta N/N$,

$$\Delta Ro/Ro = \epsilon(Ro; P)\Delta P/P + \epsilon(Ro; N)\Delta N/N$$

weighted by the system's runoff elasticities $\epsilon(Ro; X) = (\Delta Ro/Ro)/(\Delta X/X)$. These, in turn, are also related to water supply and demand and add up to one, $\epsilon(Ro; P) + \epsilon(Ro; N) = 1$. Now, simulated elasticities for changing climates are interpreted by the ideal rainfall-runoff chain. For example, at fixed water demand (or energy supply, N=constant), model output under a climate change scenario with changing precipitation only, ΔP (e.g. a shift of the monsoonal circulation), provides estimates of the rainfall dependent runoff elasticity. That is, GCM simulations determine the elasticity $\epsilon(Ro; P) = (\Delta Ro/Ro)/(\Delta P/P)$, which needs to be compared with the ideal runoff elasticity, $\epsilon(Ro; P) = 1 + D^*$, (from Eq. 3). The latter depends only on the dryness $D^* = -\ln(W)$ or runoff ratio relation $W = Ro/P$, provided by the control experiment.
Consequences are as follows: (*i*) Fig. 5 shows this runoff elasticity for the GCM's global change environment (dots) and for the ideal rainfall-runoff chain (dashed line). Note that the majority of the GCM gridpoint elasticities follows the ideal slope but with an additional offset (that reduces the elasticity by about -0.5 to -1). That is, dryness related GCM runoff elasticity $\epsilon(Ro, P) = (\Delta Ro/\Delta P)/W$ underestimates that of the ideal rainfall-runoff chain. This indicates long term memory of soil moisture storage when precipitation change is associated with a reduced runoff change. In addition, negative values of change, $(\Delta Ro/\Delta P) < 0$, may occur in areas where evaporation (exceeding precipitation) is supported by an external water supply, which are also subject of climate change. (*ii*) Note that net radiation dependent runoff elasticity can also be deduced, as $\epsilon^*(Ro, N) = -D^*$ satisfies the balance $\epsilon^*(Ro, P) + \epsilon^*(Ro, N) = 1$, if N and P are uncorrelated.

Attribution: Further insight into the rainfall-runoff processes is gained through analyzing external climate forcing and internal (say human induced) basin change by extending the classical Budyko framework to state space trajectories (following Milne et al., 2001; Tomer and Schilling, 2009; Renner et al., 2012; Cai et al., 2015). That is, utilizing relative excess rainfall and energy changes, dW and dU, which separate the water and energy related contributions, one obtains estimates of change attribution. This is measured by the attribution-ratio (dU/dW) which, for the idealized rainfall-runoff chain (see Figs. 1b and 6a), yields $(dU/dW) \sim -1$ in the dryness range, $0 < D < 2$ to 3, from Tundra via Forests and Steppe to Semi-deserts. Under the ideal conditions of watersheds in equilibrium, the following "Gedanken" experiment can be made to relate the trajecto-

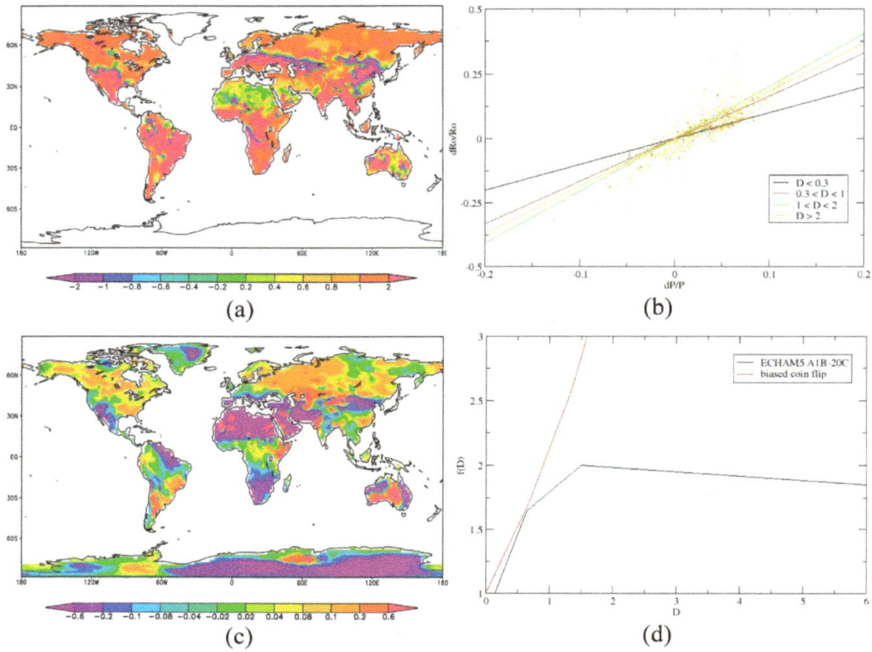

Figure 5. Analysis of rainfall induced runoff elasticity $\epsilon(Ro, P)$ based on simulations by ECHAM5 (20C and A1B-scenario): a) $\epsilon(Ro, P)$ and b) $\Delta Ro/Ro$ vs. $\Delta P/P$ grouped and regressed (the slopes of the four regression lines indicate the elasticity) over dryness D (scatter plot). c) Rainfall and potential evapotranspiration N induced relative runoff change, dRo/Ro, and d) elasticity $\epsilon(Ro, P)$ by regression $\Delta Ro/Ro = \epsilon(Ro, P) \cdot (\Delta P/P)$ over D classes and runoff elasticitiy ϵ^* based on Schreiber's formula (red line).

ries (dU, dW) of change to the attribution of the causes of the change in (U, W)-state space:

(i) Assume constant climate forcing (P and N constant) but evapotranspiration E having changed over time: then, $dU = -dE/N$ and $dW = -dE/P$ represent an internal change in flux partitioning (say dE) affecting the watershed, such as a change in vegetation or land surface, which results in a deviation from the positive diagonal (Fig. 6b, pink first quadrant and light blue third quadrant).

(ii) On the other hand, assume that E is constant; then $dD > 0$ (< 0) leads to increasing (or decreasing) $U = 1 - E/N$ and decreasing (or increasing $W = 1 - E/P$). That is, climate induced change tends along the negative diagonal toward the yellow second quadrant and dark blue fourth quadrant.

(iii) Finally, the change can be partitioned into four quadrants (Fig. 6a) and be displayed geographically (Fig. 6b) to associate regions of change with the attributed causes. For the Tibetan Plateau a first and second period (1982-93 and 1994-2006) are selected for the attribution analysis and geographical distributions of significant areas of (U, W) change exceeding $std(U)$ or $std(W)$ are displayed.

Changes in Tibet are predominantly affected by external processes (dark blue fourth quadrant: increasing W and decreasing U, or decreasing aridity). That is, wet tendencies in Tibet control vegetation dynamics in most regions, although overgrazing combined with small mammal outbreaks are considered as primary cause of increased degradation of alpine meadows. Internal (anthropogenic) effects are detected in regions (light blue third quadrant) where vegetation and excess water and energy decrease. A possible cause is related to population density increasing by about half a million. Spatial overlap between permafrost and external change-controlled regions (dark blue fourth quadrant) indicates a significant influence. Of the 21% of Tibet affected by significant (U, W)-change, 70% is attributed to external causes and 30% to internal. Note that these results refer to large-scale (U, W)-changes with related area-averaged NDVI; they compare well with Chen et al. (2014) analyzing the impact of climate change and anthropogenic activities on alpine grassland over Tibet based on a net primary production model for grassland: averaged from 1982 to 2011, about 68.5% (31.5%) of the area of actual grassland change is attributed to a changing climate (human activities). More details are described in Cai et al. (2015).

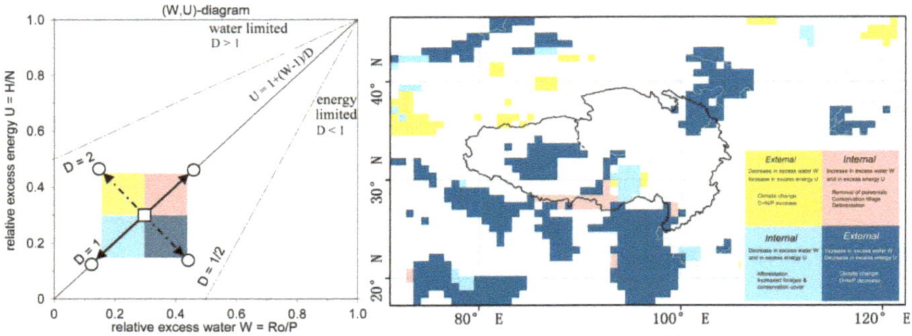

(a) (b)

Figure 6. Attribution analysis and results for Tibetan Plateau (from Cai et al., 2014): (a) Eco-hydrological states in the (U, W) diagram spanned by relative excess water W (runoff vs precipitation) and energy U (sensible heat flux vs net radiation) with lines of constant dryness D (net radiation vs precipitation). Eco-hydrological states are denoted by squares and circles: The reference states (squares) represent the first period, which is followed by subsequent shift to the second period (circles). Directions and lengths of arrows connecting first with second period provide the attribution of change: Trajectories along (across) the main diagonal characterize a change of the internal (external) partitioning. (b) Geographical distributions of attribution classes of eco-hydrological change on the Tibetan Plateau separating internal from external causes: Geographical distributions of all significant areas of (U, W) change exceeding $std(U)$ or $std(W)$.

4 Geomorphological patterns

Lakes are patterns of the Earth's surface which, jointly with drainage densities of river systems, vegetation and soil moisture characterize regional climates. In particular, terminal (closed or endorheic) lakes which, embedded in a closed basin, describe the hydrological relation between geomorphological structure of the land-lake topography, the geobotanical properties of the land-vegetation-climate system and the large-scale atmospheric conditions. In this sense, lake-basin area ratios (as a geomorphological measure) and vegetation classes (a geobotanic quantity) represent climate variables corroborating response and feedback to the atmosphere's forcing of the global circulation (in terms of the net incoming radiation and precipitation) and to the Earth's climate system.

4.1 Terminal lakes

The climatological relevance of the dryness index and its threshold, $D = 1$, separating water and energy limited climates, is demonstrated by another regional surface climate indicator as follows.

Lake area ratio: The terminal (closed or endorheic) lake area ratio combines the areas of lake, a_{lake} and watershed, a_{land}, to obtain

$$A = a_{lake}/(a_{lake} + a_{land})$$

Lake overflow occurs at $A > 1$, when $a_{land} = 0$. An equilibrium model of terminal lakes can be derived from the lake area averaged water fluxes, $P_{lake} = Ro_{lake} + E_{lake}$. Lake properties, like precipitation and runoff (= lake inflow) are indicated by subscripts. An area ratio A, which depends on basin and on lake evaporation, E and E_{lake}, and on the common precipitation, P, can be derived. Combining with the land surface water balance (Eq. 1) after weighting by the land and lake areas, and further assumptions (see Fraedrich, 2015a, for more details on this subsection): $A = (P-E)/(E_{lake} - E)$.

Minimalist model: Water balance and equation of state (Eqs. 1 and 3) determine the lake-land equilibrium climate state. This approach has been suggested by Kutzbach (1980) and successfully employed to paleo-lakes since. Instead of parameterizing land and lake radiative fluxes and the related Bowen ratios, which requires a large set of parameters, we assume the potential evaporation of lake and catchment to be of the same magnitude, $E_{lake} = PE = N$. Thus only land information is required to determine the lake's structural behavior. The lake area ratio can now be reformulated employing the equation of state. This leads to a minimalist model of terminal lakes which depends only on the dryness ratio $D = N/P$, and thus on the geobotanic state of the lake's environment (see Fig. 1a):

$$A^* = W/(D - 1 + W) = \exp(-D)/[D - 1 + \exp(-D)]$$

The star superscript is used to identify the approximation induced by the equation of state (Eq. 3). Given the relative lake area A^*, the dryness ratio D can be determined as all other flux ratios. With E_{lake} being parameterized by the potential evaporation $PE = N$ of the catchment basin, the maximum possible terminal lake area ratio is attained at $A = 1$, when the dryness ratio yields the threshold value at $D = 1$, which separates water from energy limited climates. This structure change occurs under the topographically idealized condition of a constant height basin boundary. Then the terminal lake reaches its natural overflow threshold at its largest possible extent or $a_{land} = 0$. That is, closed or terminal lakes, $0 < A < 1$, exist only under water limited land conditions, $1 < D < \infty$ and the lake area ratio decreases for increasing dryness ratio D. In contrast, overflow occurs under energy limited regimes (and ideal topography) for $D < 1$ (see Fig. 1).

Qinghai Lake – a validation analysis: Qinghai Lake (Koko Nor) is situated in the cold and arid climate of the north-eastern Qinghai-Tibetan plateau (37°N, 3200 m). It is one of China's largest closed-basin lakes, which is affected by the major Asian circulation systems: the East and SW-Asian monsoon in summer and the dry westerlies of northern Eurasia. The lake is considered to be sensitive to climate change with its area almost doubling between the extremes from present day to maximum extent during earlier climate states in the Holocene (10 to 5 kyrs bp), $a_{lake} \sim 4{,}300$ to 7,655 km^2, which is embedded in a basin area of $a_{lake} + a_{land} \sim 29 \cdot 10^3$ km^2. This corresponds to a lake area ratio ranging from $A \sim 0.145$ to $A_{max} \sim 0.258$, or interval $\delta A \sim 0.113$ (Rhode et al., 2010).
Water flux diagram: All land and terminal lake water fluxes (y-axis) depending on rainfall (x-axis) can be presented in a diagram (Fraedrich, 2015a, Figure 2; see also Fig. 7) calibrated by the dryness ratio D of the drainage area, which is determined by the lake area ratio A. The main diagonal or (1 : 1) line separates water fluxes on land (below) and lake (above the diagonal). Given the Qinghai Lake area ratio $A \sim 0.145$ (and thus the related dryness or runoff ratios, $D \sim 1.89$ or $W \sim 0.15$) and the rainfall $P \sim 0.36$ m/yr, one obtains the following results: (*i*) Land evaporation (below the main diagonal) $E = P(1 - W) \sim 0.31$ m/yr and runoff $Ro = PW \sim 0.05$ m/yr add up to the rainfall $P \sim 0.36$ m/yr (main diagonal). (*ii*) Lake evaporation (above the main diagonal), $E_{lake} = DP \sim 0.68$ m/yr (steepest line) is balanced by rainfall $P = P_{lake}$ (main diagonal) plus inflow from the catchment, $Ro_{lake} \sim 0.32$ m/yr. The latter agrees with the measured runoff from the drainage area, $Ro_{land}(1 - 1/A) \sim 0.32$ m/yr, because land runoff and lake inflow balance, $a_{land}Ro_{land} + a_{lake}Ro_{lake} = 0$. Note that, for a closed catchment-terminal lake system in equilibrium, the total rainfall balances the total evaporation at the surface and there is no net divergence or convergence of atmospheric or subsurface water.

Lake Chad – a sensitivity analysis: The Lake Chad basin covers an area of about 2.5 million km^2, which is situated in Northern Africa (6 to 24°N, 7 to 24°E). Its climatological setting shows a large meridional gradient of rainfall with 1.6 m/yr and larger (0.15 m/yr and smaller) in the

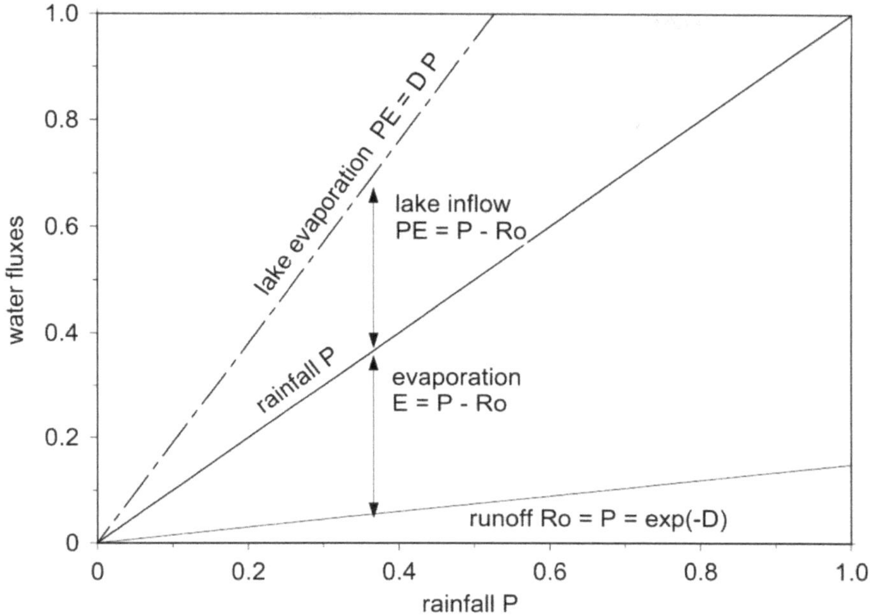

Figure 7. Water fluxes derived from the minimalist model of terminal Lake Qinghai with lake area ratio $A \sim 0.145$ dependent dryness ratio, $D \sim 1.9$ and the rainfall $P \sim 0.36$ m/yr. (i) Land water fluxes comprise evaporation E and runoff Ro adding up to precipitation P (main diagonal). (ii) Lake evaporation $PE = E_{lake} = DP$ for $D > 1$ exceeds the main diagonal $(1:1)$ because it is balancing the sum of precipitation P (main diagonal) and inflow (Ri) from the catchment, $Ro_{lake} = P - PE = (1 - D)P < 0$.

southern (northern) regions dominated by the West-African monsoonal circulation. The vegetation is characterized by desert and steppe in the north, savanna and woodland in the south. The complex structure of the Lake Chad basin allows the classification of present and paleo-lake state related to topographical thresholds.

(i) Validation and Normal Chad: The water budget of Normal Lake Chad (1954-1969) has been extensively analyzed (see Fraedrich, 2015a) and serves as reference for validating the minimalist model. The Normal Lake Chad area ratio, $A_0 \sim 0.008$ corresponds to the dryness and runoff ratios $(D, W)_0 \sim (3.6, 0.027)$. Given the mean annual pre-1970 runoff, $Ro \sim 0.017$ m/yr, we obtain minimalist model estimates of catchment mean precipitation

($P = Ro/W$), $P \sim 0.63$ m/yr, and lake evaporation ($E_{lake} = PE = DP$), $PE \sim 2.3$ m/yr. These water fluxes correspond surprisingly well with the observed data. The annual average zonal mean precipitation over Africa from 6 to 24°N, 7 to 24°E (meridional extent of Lake Chad catchment) decreases from about 1.3 to 0.05 m/yr or less (see Krinner et al., 2012, Figure 1) leading to an estimate of an average of about 0.65 m/yr. This validated dataset provides the reference state, A_0 and P_0, for the subsequent sensitivity analysis, to estimate the change of area ratio, δA, from present day to Mega-Chad conditions using a partially linear tangent approach of the minimalist equilibrium model.

(ii) Sensitivity and Mega-Chad: Under water limited conditions, $D > 1$, a sensitivity analysis of the reference climate state (subscript 0) is expressed in terms of dryness ratio variations:

$$\delta D/D_0 = \delta PE/PE_0 - \delta P/P_0$$

For dryness sensitivity to be negative (left), the increase in rainfall needs to be larger than the increase of potential evaporation. Under water limited or dry climate conditions, $D > 1$, a changing dryness ratio, $\delta D = (\delta PE - D_0 \delta P)/P_0$, shows that only a small amount of rainfall increase is required to shift the dryness to a wetter state (and vice versa): $\delta PE \ll D_0 \delta P$. Thus, in particular, the sensitivity of dry (steppe and savanna) and very dry (semi-desert and desert) climates is strongly affected by changes in rainfall, even if these changes are rather small; that is, $\delta D \sim -D_0 \delta P/P_0$, or $D \sim D_0(1 - \delta P/P_0)$, which gives the dryness changing with rainfall or vice versa, $D \sim D_0(2 - P/P_0)$ or $P \sim P_0(2 - D/D_0)$. Now, Mega-Chad (subscript "mega") is characterized by a substantially larger area ratio $A_{mega} \sim 0.14$ and wetter climate with dryness and runoff ratios $(D, W)_{mega} \sim (1.90, 0.15)$ deduced by the minimalist model. Our sensitivity analysis yields a rainfall estimate of $P_{mega} \sim P_0(2 - D_{mega}/D_0) \sim P_0(2 - 1.9/3.6) \sim 0.93$ m/yr (see Krinner et al., 2012, Figure 1). In the Holocene, the annual average zonal mean precipitation over Africa between 6 and 24°N (meridional extent of Lake Chad catchment) has been simulated to increase by about 1.6 and 0.2 m/yr leading to an average of about 0.9 m/yr. "The relative precipitation increase" (compared to the control run) is particularly strong, in excess of 50%, in the central and western Sahara, where the annual mean precipitation rates are about 0.25 m/yr. In summary, the minimalist model of terminal lakes provides sensitivity estimates for past climates without requiring the large number of land and/or lake surface parameters (such as albedo, net radiation, Bowen ratio, but only two, which characterize the reference climate (subscript "0"): Here rainfall P_0 and lake area ratio A_0 (or dryness D_0) refer to present day climate. Note that our minimalist model has been successfully applied to lake Eyre in Australia (Larsen, 2011).

4.2 Fractional soil moisture

When long time scales are analyzed, the chaotic nature of the water cycle (including soil water storage change dV/dt) is averaged out and only residuals emerge so that surface energy and water budgets account for most of the responses. However, the soil water volume, $V = ZS$, is relevant as a climate mean state variable, because it describes the vertically averaged relative soil-moisture S, which is related to an effective (with respect to soil porosity) soil water bucket model with (root) depth Z, and to the evaporation process

$$S = E/N$$

Employing the equation of state, one obtains $S^* = 1 - U = 1 - \exp(-D)/D$ (see Fig. 1a). It appears (Fig. 8) that the bucket model for soil moisture (based on the Schreiber formula, Eq. 3) is a useful soil moisture indicator for watershed scale, which depends on dryness as the only climate indicator.

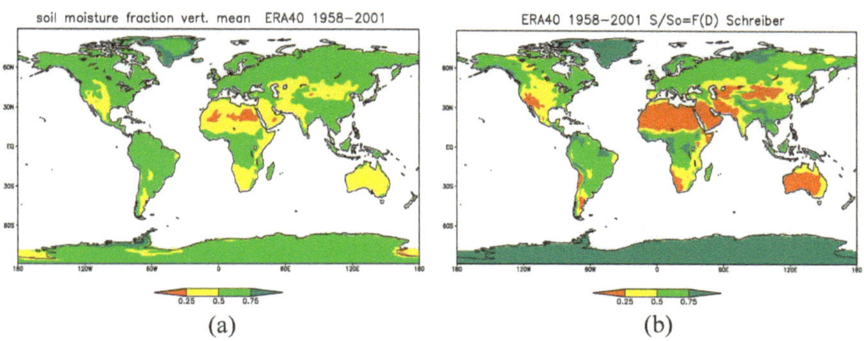

Figure 8. Fractional soil moisture from ERA-40 data (1958 to 2001): (a) vertical mean and (b) soil moisture bucket, $S^* = E/N = 1 - \exp(-D)/D$, based on the equation of state (Eq. 3) and dryness D.

4.3 Vegetation: Dryness and NDVI climate and variability

Before demonstrating the seasonal and interannual variability the geographical distribution of the long term means of (a) energy demand versus water supply (or dryness index D) from ERA-Interim and (b) vegetation from NDVI datasets are presented (Fig. 9). The first enters the analysis in terms of the dryness ratio, $D = N/P$, to measure the climate's water (or energy) limitation; at $D = 1$ water and energy limitation separated. According to Budyko (1974), dryness is divided into five biome ranges (see Fig. 2, D=0–0.3, 0.3–1, 1–2, 2–3, and >3). Vegetation is presented by NDVI-greenness (Weier and Herring, 2005; Wittich and Hansing, 1995): very low values (0.1 and below corresponding to barren areas, sand, or snow) plus moderate values (0.2–0.3 as shrub and grassland), and high values (0.6–0.8 for temperate and tropical rainforests) are combined so that three ranges are obtained: 0.1 – 0.3 (grey), 0.3 – 0.6 (brown), and 0.6 – 1 (green).

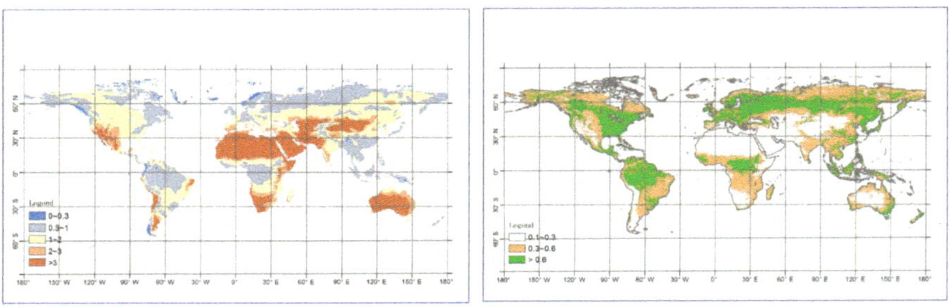

(a) (b)

Figure 9. Geographical distributions of a) Budyko's dryness index and b) NDVI greenness index.

Seasonal variability: The interaction between dryness D and NDVI (Fig. 10) indicates a significant seasonality. Previous works about NDVI show that very low values of NDVI (0.1 and below) correspond to barren areas, sand, or snow. Therefore, in annual seasonal effect analysis, we did not consider those non-vegetation types, but focus on the relationship among vegetation growth (0.1–1), water availability and solar radiation intensity. Vegetation (NDVI in the classes 0.2–0.3 and 0.6–0.8) shows more seasonal dependency compared to the moderate values representing shrub and grassland (0.2–0.3), while high values indicate temperate and tropical rainforests (0.6–0.8). Thus, the seasonal dependency proves that, although temperate and tropical rainforests are under energy-limited conditions in non-summer seasons, they grow because of sufficient water supply. That is when energy increases in summer, their growth increases as well.

Interannual variability: The combination of ERA-Interim climate and GIMMS NDVI vegetation information in the (N, D) diagram provides a

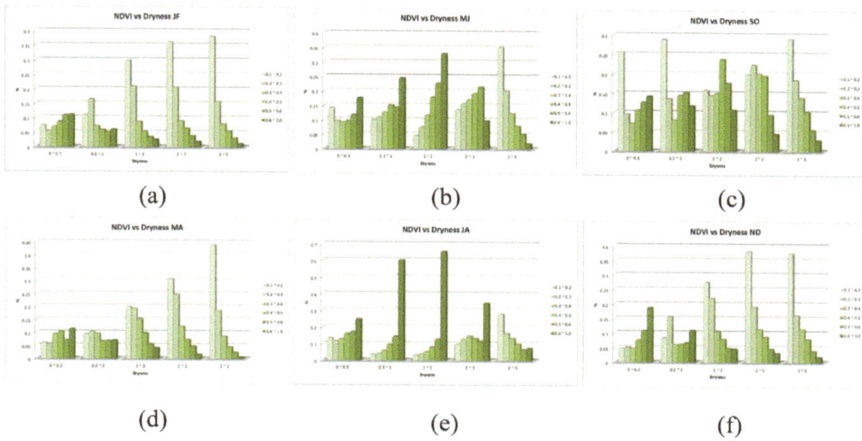

Figure 10. Temporal distributions of bi-monthly Vegetation Index: NDVI grouped by dryness index. JF represents NDVI mean from January to February, similarly, MA = March April, MJ = May June, JA = July August, SO = September October, ND = November December.

suitable background to extend the analysis beyond the first moments and to include interannual variability. The coefficient of variation as the ratio of standard deviation to the mean (or inverse of signal to noise ratio) is used to compare the measure of dryness (ERA-Interim) and NDVI. A small coefficient of dryness variation demonstrates that high values of climate mean NDVI (straddling $D = 1$) favor regions with small dryness variability or, vice versa, the green vegetation (associated with forests) leads to small variations of dryness. The median of the coefficients of variation binned in NDVI classes of greenness (width of 0.1) clearly decrease with increasing dryness. Apparently vegetation greenness tends to smooth the volatility of the year-to-year fluctuations of dryness. That is, large NDVI greenness of vegetation straddles the water-energy limitation (at $D = 1$) where also minimum values of the coefficient of year-to-year dryness variation are attained. The dryness variation versus dryness mean boxplot supports the analysis. Note that bins in Fig. 11c are chosen according to Budyko's D-classes. Values for desert conditions are grouped in a single class, $D > 3$.

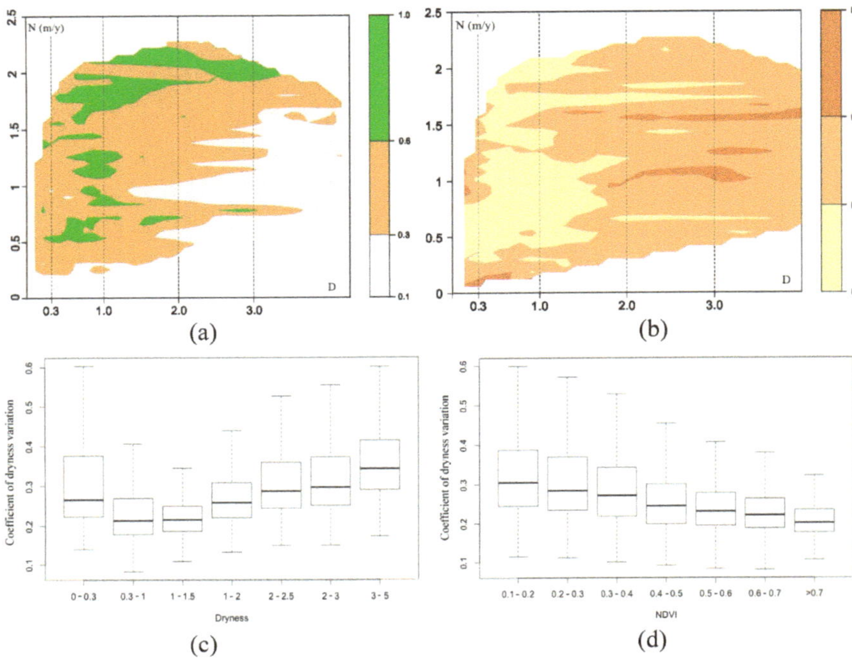

Figure 11. Vegetation analysis in the net radiation dryness (N, D) diagram (units in m/yr water equivalent; smoothing by linear interpolation): (a) Northern Hemisphere NDVI-greenness and (b) coefficient of year-to-year dryness variation (ratio of standard deviation to climate mean). Box plots present the statistics of the coefficients of dryness variation depending on (c) mean NDVI greenness vegetation and (d) mean dryness D.

5 Summary and Outlook: On the human dimension – risk state space

The processes along the rainfall-runoff chain, which characterize the Earth surface, are analyzed in the Budyko framework using a set of non-dimensional water and energy flux ratios spanning eco-hydrological state spaces. First, the ideal rainfall-runoff chain is introduced communicating between water supply and demand and satisfying water and energy budgets and an equation of state. Secondly, various eco-hydrological state spaces, their variables and diagrams are presented including first analyses of geobotanic classifications, a validation of the parsimonious concept of the rainfall-runoff chain.

The application to change is presented in terms of sensitivity (or elasticity) analyses and a diagnostic of change attribution to internal and external causes, that is climate and human induced. Finally, geomorphological patterns related to vegetation, rivers, soil, and lakes are analyzed and conceptual modeling following the parsimonious approach is applied. Here in the conclusion a brief outlook to the human dimension is given. It is associated with the risk concept, by which the diagnostics of damage and its circumstances turns into a prognosis provided by an estimate of the probabilities of its occurrence. Linking physical and human dimension is obtained by a two-dimensional space of risks spanned by damage versus probability of occurrence, in which three regions are identified: Normal, transition and prohibited areas. Adding confidence intervals to the damage and the probability estimates enlarges the two dimensional phase space of Global Change risks, whose dimension can be further increased by including information on their spatial extent, their duration and type of evolution, that is the ubiquity, persistence and irreversibility of the global risks. In this effectively higher dimensional phase space risk classes are formed, which may contain risks of different origin. Seven global risk classes have been identified which, except for the unknown risk, are coined after the Greek mythology: Cyclops, Cassandra, Medusa, Pandora, Damocles, Phythia (Beese et al., 1998, Figures A2-2, A2-3; see also Fig. 12). For example: Cyclops are mighty giants who, for all their strength but being one-eyed, can only perceive one side of reality. Thus, Cyclops class risks are characterized by a probability of occurrence, which is highly uncertain but, if damage occurs, it is very large. Many natural events like floods and earthquakes fall in this class. The high uncertainty of occurrence requires three strategies: ascertaining the probability, prevent surprises, and manage emergency. This introduction of the human dimension, associated with predictability or uncertainty estimates of Global Change may suffice as a snapshot of a research area linking the social, economic, and natural science communities but it needs to be extended to regional risks on watershed scale.

Acknowledgments

We thank Dr. Elisa Palazzi, Institute of Atmospheric Sciences and Climate (ISAC), National Research Council (CNR) for her constructive comments on this lecture note.

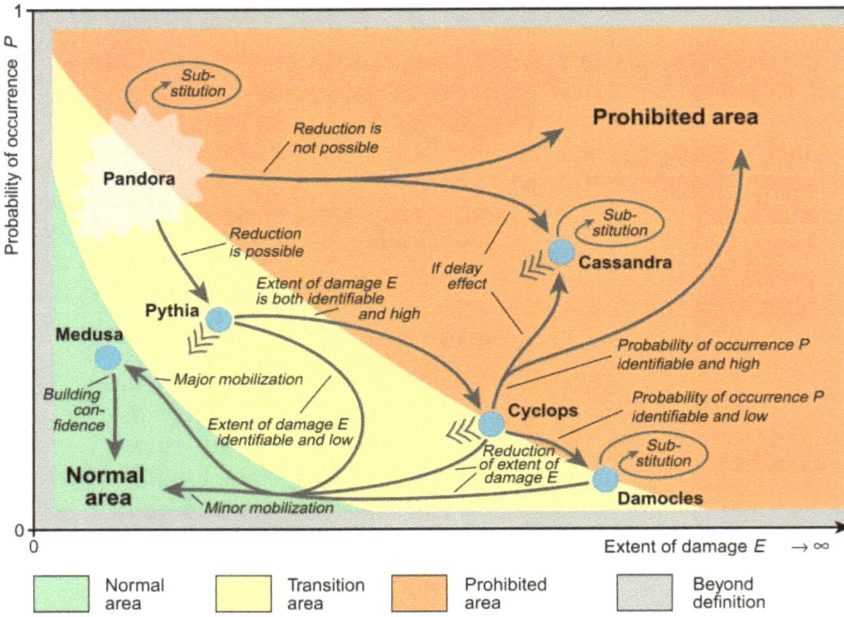

Figure 12. Risk Classes of Global Change: Their location in the normal, transition and prohibited regions of the (damage, probability)-space. Action of society may move a risk from one region to another, even by perception. A concerted action, however, is required, if risks are to be reduced and shifted to the normal region. Therefore, each risk class requires its own reduction strategy, which is related to its location in phase space and the uncertainty of its estimate (from Beese et al., 1998, Figure A4.1-1).

Bibliography

Beese, F. O., K. Fraedrich, P. Klemmer, J. Kokott, L. Kruse-Graumann, C. Neumann, O. Renn, H. -J. Schellnhuber, E. -D. Schulze, M. Tilzer, P. Velsinger, H. Zimmermann, *World in Transition: Strategies for Managing Global Environmental Risks*, Berlin, Heidelberg, New-York: Springer, 383pp, 1998.

Budyko, M. I., *Climate and Life*, Academic Press, New York, 508 pp., 1974.

Cai, D., K. Fraedrich, F. Sielmann, Y. Guan, S. Guo, L. Zhang, X. Zhu, *Climate and vegetation: An ERA-1 Interim and GIMMS NDVI Analysis*, J. Climate, **27**, 5111-5118, doi: 10.1175/JCLI-D-13-00674.1, 2014.

Cai, D, K. Fraedrich, F. Sielmann, L. Zhang, X. Zhu, S. Guo, Y. Guan, *Vegetation dynamics on the Tibetan Plateau (1982 to 2006): An attribution by eco-hydrological diagnostics*, J. Climate, **28**, 4576-4584, doi: 10.1175/JCLI-D-14-00692.1, 2015.

Chen, B., X. Zhang, J. Tao, J. Wu, J. Wang, P. Shi, Y. Zhang, C. Yu, *The impact of climate change and anthropogenic activities on alpine grassland over the Qinghai-Tibet Plateau*, Agric. For. Meteor., **189-190**, 11-18, doi:10.1016/ j.agrformet.2014.01.002, 2014

Fraedrich, K., *Predictability: Short- and long term memory of the atmosphere.* In Chaos in Geophysical Flows (eds. G. Boffetta, S. Larcotta, G. Visconti and A. Vulpiani). International Summer School on Atmospheric and Oceanic Sciences (ISSAOS 2001), Otto Editore, Torino, Italy, 63-104, 2005.

Fraedrich, K., A. A. Aigner, E. Kirk, F. Lunkeit, *General Circulation Models of the Atmosphere.* In Encyclopedia of Nonlinear Science, Routledge, New York, 359-361, ISBN: 1-57958-385-7, 2005

Fraedrich, K., *Forecasting: Predictability, Probability, and Persistence*, Mitt. Math. Ges. Hamburg, **26**, 5-28, 2007.

Fraedrich, K., *A stochastic parsimonious water reservoir: Schreiber's 1904 equation* J. Hydrometeor., **11**, 575-578, doi: 10.1175/2009JHM1179.1, 2010.

Fraedrich, K., F. Sielmann, *An equation of state for land surface climates*, Int. J. Bifurcation Chaos, **21**, 3577-3587, doi: 10.1142/S021812741103074X, 2011.

Fraedrich, K., F. Sielmann, L. Zhang, X. Zhu 2013, *Applications of an ideal rainfall-runoff chain: validation and climate change* European Water Resources Association (EWRA), 845-852, 2013.

Fraedrich, K., *A minimalist model of terminal lakes: Qinghai Lake (China) and Lake Chad (N-Africa)*, Hydrology Research, **46**, 222-231, 2015a.

Fraedrich K., F. Sielmann, D. Cai, L. Zhang, X. Zhu *Validation of an ideal rainfall-runoff chain in a GCM environment*, Water Resources Management, **29**, 313-324, doi: 10.1007/s11269-014-0703-2, 2015b

Fraedrich, K., I. Bordi, X. Zhu, *Climate dynamics on global scale: resilience, hysteresis and attribution of change.* In: The Fluid Dynamics of Climate, CISM, Springer Verlag., 2015c

Hanasaki N., S. Kanae, T. Oke, K. Masuda, K. Motoya, N. Shirakawa, Y. Shen, K. Tanaka, *An integrated model for the assessment of global water resources – Part 1: model description and input meteorological forcing*, Hydrol. Earth. Syst. Sci., **12**, 1007-1025, doi:10.5194/hess-12-1007-2008, 2008.

Holden, P. B., N. R. Edwards, P. H. Garthwaite, K. Fraedrich, F. Lunkeit, E. Kirk, M. Labriet, A. Kanudia, F. Babonneau, *PLASIM-ENTSem: a spatio-temporal emulator of future climate change for impacts assessment*, Geosci. Model Dev., **7**, 433-451, doi:10.5194/gmd-7-433-2014, 2014.

Koeppen, W., *Das geographische System der Klimate – Handbuch der Klimatologie*, vol. 1, PartC. Gebruder Borntraeger, Berlin, 388, 1936.

Krinner, G., A.- M. Lezine, P. Bracconot, P. Sepulchre, G. Ramstein, C. Grenier, I. Gouttevin, I., *A reassessment of lake and wetland feedbacks on the North African Holocene climate* Geophys. Res. Lett., **39**, L07701, doi: 10.1029/2012GL050992, 2012.

Kutzbach, J. E. *Estimates of past climate at paleolake Chad, North Africa, based on a hydrological and energy balance model*, Quaternary Res., **14**, 210-223, doi: doi:10.1016/0033-5894(80)90049-6, 1980.

Larsen, J. R., *Aspects of the contemporary and Quaternary hydrology of the Lake Eyre Basin, central Australia*, Doctor of Philosophy Thesis, School of Earth and Environmental Sciences, University of Wollongong, http://ro.uow.edu.au/thesis/3538, 2011.

Milne, B., V. Gupta, C. Restrepo, *A scale invariant coupling of plants, water, energy, and terrain*, Ecoscience, **9**, 191-199, 2002.

Popper, K. R., *Of Clocks and Clouds*, Objective Knowledge. Oxford, At the Clarendon Press, 206-255, 1972.

Renner, M., R. Seppelt, C. Bernhofer, *Evaluation of water-energy balance frameworks to predict the sensitivity of streamflow to climate change*, Hydrol. Earth Syst. Sci., **16**, 1419-1433, doi:10.5194/hess-16-1419-2012, 2012.

Rhode, D., H. Ma, D. B. Madsen, P. J. Brantingham, S. L. Forman, J. W. Olsen, *Paleo-environmental and archaeological investigations at Qinghai Lake (western China): Geomorphic and chronometric evidence of lake level history*, Quaternary Int., **218**, 29-44, doi: doi:10.1016/j.quaint.2009.03.004, 2010.

Roeckner E., M. Lautenschlager, H. Schneider, *IPCC-AR4MPI-ECHAM5-T63L31MPI-OMGR1.5 L40SRESA1B run no. 1: atmosphere monthly mean valuesMPImet/MaD Germany*, World Data Cent Clim, oi:10.1594/WDCC/EH5- T63L31-OM-GR1.5L40-A1B-1-MM, 2006.

Schreiber, P., *Über die Beziehungen zwischen dem Niederschlag und der Wasserfuhrung der Flusse in Mitteleuropa [On the relation of rainfall and streamflow of central Europe]*, Meteorolog. Z., **21**, 441-452, 1904.

Stephenson, N. L., *Climatic control of vegetation distribution: The role of the water balance*, American Naturalist, **135**, 649-670, 1990.

Tomer, M. D., K. E. Schilling, *A simple approach to distinguish land-use and climate-change effects on watershed hydrology*, J. Hydrol., **376**, 24-33, doi:10.1016/j.jhydrol.2009.07.029., 2009

Weier, J., D. Herring, *Measuring vegetation (NDVI and EVI)* Earth Observatory Library of NASA, http://earthobservatory.nasa.gov/Library/MeasuringVegetation/, 2005.

Wittich, K., O. Hansing, *Area-averaged vegetative cover fraction estimated from satellite data*, Int. J. Biometeorol., **38**, 209-215, doi: 10.1007/BF01245391, 1995.

FSC
www.fsc.org
MIX
Papier | Fördert gute Waldnutzung
FSC® C083411

Kolibri 360 GmbH
Ferdinand-Jühlke-Straße 7
99095 Erfurt, Deutschland
produktsicherheit@kolibri360.de